P9-EBY-593

# INDUSTRIAL ENVIRONMENTAL PERFORMANCE METRICS

## CHALLENGES AND OPPORTUNITIES

Committee on Industrial Environmental Performance Metrics

NATIONAL ACADEMY OF ENGINEERING
NATIONAL RESEARCH COUNCIL

NATIONAL ACADEMY PRESS
Washington, D.C.

**NATIONAL ACADEMY PRESS • 2101 Constitution Avenue, N.W. • Washington, D.C. 20418**

NOTICE: The project that is the subject of this report was approved by the Governing Board of the National Research Council, whose members are drawn from the councils of the National Academy of Sciences, the National Academy of Engineering, and the Institute of Medicine. The members of the committee responsible for the report were chosen for their unique expertise and with regard for appropriate balance.

The National Academy of Engineering was established in 1964, under the charter of the National Academy of Sciences, as a parallel organization of outstanding engineers. It is autonomous in its administration and in the selection of its members, sharing with the National Academy of Sciences the responsibility of advising the federal government. The National Academy of Engineering also sponsors engineering programs aimed at meeting national needs, encourages education and research, and recognized the superior achievements of engineers. Wm. A. Wulf is the president of the National Academy of Engineering.

Support for this project was provided by the United States-Asia Environmental Partnership, a program of the Agency for International Development (under grant no. AEP-A-00-97-00014-00). The views presented in this report are those of the Committee on Industrial Environmental Performance Metrics and are not necessarily those of the funding organization.

**Library of Congress Cataloging-in-Publication Data**

Industrial environmental performance metrics : challenges and opportunities / Committee on Industrial Environmental Performance Metrics, National Academy of Engineering, National Research Council.
    p. cm.
    Includes bibliographical references and index.
    ISBN 0-309-06242-X (pbk. : alk. paper)
    1. Production management—Environmental aspects. 2. Industrial ecology—United States—Measurement. 3. Environmental policy—United States. I. National Research Council (U.S.). Committee on Industrial Environmental Performance Metrics.
    TS155.7 .I5 1999
    658.4'08—dc21                 99-6298
                                   CIP

ISBN 0-309-06242-X

Cover art: *Like Wild Horses* (detail), courtesy of the artist, Kay Jackson, Washington, D.C.

Copyright 1999 by the National Academy of Sciences. All rights reserved.

This book is printed on recycled paper.

Printed in the United States of America

## COMMITTEE ON
## INDUSTRIAL ENVIRONMENTAL PERFORMANCE METRICS

ROBERT A. FROSCH (*Chair*), Senior Research Fellow, Center for Science
and International Affairs, John F. Kennedy School of Government,
Harvard University
DAVID C. BONNER, Director of Technology, Performance Polymers
Division, Rohm and Haas
JOHN B. CARBERRY, Director, Environmental Technology, E.I. DuPont
LESLIE CAROTHERS, Vice President, Environment, Health, and Safety,
United Technologies Corporation
DARYL DITZ, Director, Environmental Management Program, Environmental
Law Institute
THOMAS N. GLADWIN, Max McGraw Professor of Sustainable Enterprise
and Director, Corporate Environmental Management Program, University
of Michigan
THOMAS E. GRAEDEL, Professor of Industrial Ecology, Professor of
Chemical Engineering, and Professor of Geology and Geophysics, School
of Forestry and Environmental Studies, Yale University
CHRISTOPHER (KIT) GREEN, Executive Director, Materials Research and
Technology Business Development Directorate and Chief Technology
Officer, China, General Motors
RICHARD R. GUSTAFSON, Denman Professor of Paper Science and
Engineering and Chair, Management and Engineering Department,
University of Washington College of Forest Resources
MICHAEL J. LEAKE, Director, Environment, Health, and Safety, Raytheon/
Texas Instruments Systems
DAVID W. MAYER, Director, Pollution Prevention and Environmental
Performance, Georgia-Pacific
RICHARD D. MORGENSTERN, Senior Fellow, Resources for the Future
WILLIAM F. POWERS, Vice President, Research, Ford Motor Company
DARRYL K. WILLIAMS, Senior Vice President, Technology, Eastman
Chemical

**Staff**

DEANNA J. RICHARDS, Study Director
GREGORY W. CHARACKLIS, NAE Fellow
GREG PEARSON, NAE Editor
LONG T. NGUYEN, Project Assistant

# Preface

It is often asserted that "what's measured gets managed." Although this principle is now widely accepted, relatively little attention has focused on examining the metrics that organizations use to measure environmental performance. Analysis of these metrics reveals much about which environmental parameters are being incorporated into decision making while also identifying those areas receiving less attention. Interest in industrial environmental performance metrics is increasing as companies find new internal uses for such information in applications such as marketing and product development. Interest is also on the rise among external stakeholder groups (e.g., customers, local communities, the financial sector), many of which are making greater use of publicly reported environmental information. External groups have integrated such information into product buying decisions, investment or lending decisions, and investigations into the environmental performance of corporations operating in their neighborhoods. The increasing environmental awareness of both industry and society at large will only increase demand for information on environmental performance. Studies of this kind serve to identify opportunities for improvement as well as challenges to be overcome along the way.

In mid-1997 the U.S.-Asia Environmental Partnership, a program of the U.S. Agency for International Development (USAID), requested that the National Academy of Engineering (NAE) undertake a study of the use of industrial environmental performance metrics in several U.S. industries. The purpose was to identify a set of metrics that would find broad utility across industries and assist in the setting of both national and industrial environmental goals. The U.S.-focused study was to be the American contribution to a larger effort by the Asia

Pacific Economic Cooperation on industrial environmental indicators and clean production.

In response to the USAID request, the NAE, in concert with the National Research Council, established the Committee on Industrial Environmental Performance Metrics. The committee was chaired by Robert A. Frosch, former vice president for research at General Motors and now senior research fellow at the Center for Science and International Affairs, John F. Kennedy School of Government, Harvard University. The committee was carefully assembled to ensure that its members, individually and collectively, had unique expertise with respect to environmental performance metrics in one of the chosen industrial sectors or in public policy applications.

The committee launched a study to gather evidence and explore present efforts in the development and use of industrial environmental performance metrics. The study investigated such issues as the motivation for improved environmental performance, normalization of metrics, standards for reporting, aggregation of metrics, and weighting of metrics. Looking ahead at the changing needs in this area, the committee also delved into the role of metrics as a tool to both drive and measure aspects of sustainable development. This report represents the committee's collective wisdom and makes an important contribution to a complex and often confusing debate over what constitutes a measure of industrial environmental performance.

On behalf of the National Academy of Engineering, I would like to thank the chair and members of the committee (p. iii) for their insights and efforts on this project. Over the course of a year, several committee meetings, one workshop, and innumerable faxes, e-mails, phone calls, and draft versions of the report, they remained actively engaged and unfailingly constructive. Critical to the success of this effort have been several uncompensated consultants: Wayne France, David Moore, Irving Salmeen, Angie Schurig, and Ronald L. Williams. Without their contributions this project would have taken twice as long and been half as productive. They all deserve special thanks.

This report has been reviewed in draft form by individuals chosen for their diverse perspectives and technical expertise, in accordance with procedures approved by the NRC's Report Review Committee. The purpose of this independent review is to provide candid and critical comments that will assist the institution in making the published report as sound as possible and to ensure that the report meets institutional standards for objectivity, evidence, and responsiveness to the study charge. The review comments and draft manuscript remain confidential to protect the integrity of the deliberative process. We wish to thank the following individuals for their participation in the review of this report:

Hamid Aarastoopour, Illinois Institute of Technology; Henry M. Conger, Homestake Mining Company; Alexander H. Flax, Consultant; Harold Forsen, National Academy of Engineering; Stuart Hart, University of

North Carolina Business School; Jeffrey S. Hsieh, Georgia Institute of Technology; William Howard, Consultant; Robert Pfahl, Motorola; Ross Stevens, Stevens Associates; and Thomas L. Theis; Clarkson University.

While the individuals listed above have provided constructive comments and suggestions, it must be emphasized that responsibility for the final content of this report rests entirely with the authoring committee and the institution.

I would also like to thank several members of the NAE Program Office. Deanna J. Richards, who directs the NAE's Technology and Sustainable Development program, served as study director. NAE Fellow Greg Characklis was deeply involved in the formulation and direction of the project from start to finish and played a critical role in managing the committee process. Long Nguyen provided critical research, administrative, logistical, and editorial assistance. Greg Pearson, the Academy's editor, contributed invaluable and steadfast editing and publishing oversight of this document.

Finally, I would like to thank the USAID, sponsor of this effort, which was generous with help and advice throughout the project.

Wm. A. Wulf
President
National Academy of Engineering

# Contents

APPENDIXES

# INDUSTRIAL ENVIRONMENTAL
# PERFORMANCE METRICS

# Executive Summary

## BACKGROUND

The ability to gauge improvement in any endeavor is critically dependent on establishing valid methods of measuring performance. Tracking progress toward an established goal serves to influence behavior by providing continual feedback, and it requires reliable and consistent metrics. In response to regulation, competitive pressures, and the increasing demands of a wide range of stakeholders, many companies have begun investigating metrics that will improve their ability to assess environmental performance.

While many industries have significantly lessened the environmental impact of their operations over the past several decades, the majority of these improvements have been driven by the need to comply with federal or state laws. Accordingly, most measures of environmental performance are based on government reporting requirements (e.g., related to the release of regulated substances or the disposal of hazardous wastes). Recently, however, an increasing number of firms have begun to view environmental performance as an area of potential competitive advantage (Fischer and Schot, 1993; KPMG, 1997; KPMG Denmark, 1997a; Porter, 1991). As industry has begun to develop environmental goals that move beyond compliance, new methods of measuring and tracking improvement have been required (Ditz and Ranganathan, 1997; White and Zinkl, 1996).

To investigate the potential for advancement in the area of industrial environmental performance metrics, the National Academy of Engineering, at the request of the U.S. Agency for International Development's U.S.-Asia Environmental Partnership, undertook a study with the following objectives:

*1*

- to examine U.S. industrial experience in setting goals and measuring progress in environmental performance;
- to identify factors that have contributed, or will contribute, to improved industrial environmental performance in the United States;
- to assess the relative successes and shortcomings of current methods of measuring industrial environmental performance; and
- to recommend a set of industrial environmental performance metrics that define current best practices and identify directions for future improvement.

This study differs from other efforts to investigate environmental metrics in two key aspects. First, it is an industry-centered analysis driven by a committee primarily composed of professionals with corporate and manufacturing experience. Second, in an attempt to provide more in-depth analysis rather than a general investigation, this effort focuses on only four major manufacturing industries: automotive, chemical, electronics, and pulp and paper. While many sectors are notably absent (e.g., services industry), the sectors examined in this study represent a significant portion of U.S. industrial capacity (12.6 percent of gross domestic product in 1995 [United States Bureau of the Census, 1997]). Operations within these industries combine to span the full spectrum of the product life cycle as well as a wide range of market niches (e.g., raw materials, component parts, finished consumer products). Some findings of this study may, therefore, be relevant to other industry sectors or even more broadly on a national level. Industry selection was also influenced by a given sector's relevance within nations of the Asia-Pacific Economic Cooperation, in keeping with the project's objective of disseminating U.S. best practices to these economies.

Throughout the study process the committee noted a changing and more proactive attitude toward corporate environmental management within many companies. Most members of the committee held the view that companies will be expected to meet ever higher environmental standards in the future. However, there was also a feeling that if these standards continue to be developed and enforced through the same traditional and often adversarial process as in the past, both economic and environmental progress will suffer. While the committee was optimistic about the ability of improved metrics to provide opportunities for profit, there was also an understanding that more transparent measures could provide companies greater ability to communicate their voluntary environmental efforts, both to the public and to regulators.

## OBSERVATIONS, TRENDS, AND CHALLENGES

Environmental metrics are at the heart of how industry and its many stakeholders define environmental performance and determine whether progress is being made. This report documents how the use of environmental metrics has focused the attention of industry, public agencies, and a variety of other interested parties on a set of key areas of performance. All four U.S. industries in this study have integrated some metrics (Table E-1), especially for pollution releases and hazardous waste generation, into routine management decisions and external reporting. In addition, many U.S. firms track their consumption of energy, water, and some material resource inputs as a basic element of cost control. More recently, some firms have begun to make use of environmental performance measures to demonstrate to the public an increased commitment to environmental stewardship.

The motivation for measuring and tracking environmental performance in these areas is relatively consistent across industries, with the drivers falling into three general categories:

- compliance with regulatory statutes,
- achievement or strengthening of competitive advantage, and
- improvement of corporate stewardship and reputation.

TABLE E-1  Environmental Metrics Used in the Four Industry Sectors

| Metric | Automotive | Chemical | Electronics | Pulp and Paper |
|---|---|---|---|---|
| Supply Chain | E | | E | |
| **Facility Centered** | | | | |
| Pollutant releases | C | C | C | C |
| Greenhouse gas emissions | C | C | C | |
| Material use | C | C | E | C |
| Percent recycled material | C | C | E | C |
| Energy use | C | C | C | C |
| Water use | C | C | C | C |
| Packaging | C | C | C | C |
| Percent of land preserved | | E | | C |
| Environmental incidence report | C | C | C | C |
| Lost workdays/injuries | C | C | C | C |
| **Product Centered** | | | | |
| Nongreenhouse gas emissions | C | | | |
| Greenhouse gas emissions | C | | | |
| Material use | | | | |
| Energy use | C | | C | |
| **Sustainability** | | | | |
| Sustainable forestry | | | | E |

NOTE: C = environmental metric in current use; E = emerging environmental metric.

Corporate attitudes toward environmental performance are evolving beyond a strict concern for compliance, as some firms have begun to realize substantial financial benefits as a result of improved environmental performance (Buzzelli, 1996, Popoff, 1993; 3M, 1998). The emergence of more comprehensive means of assessing corporate costs, as described by the principles of full-cost accounting, has demonstrated that current procedures often underrepresent the magnitude of environmental costs (Ditz et al., 1995; Epstein, 1996; Popoff, 1993; United States Environmental Protection Agency, 1995a, 1995b). Corporate environmental information is also being increasingly sought by commercial stakeholders (e.g., lenders, insurers, investors), particularly in response to concerns over liability. Finally, a growing number of companies are making use of environmental metrics as a tool to communicate improved environmental performance to local communities and regulatory bodies.

With new knowledge and changing public expectations, fresh environmental challenges are arising that are not addressed by contemporary environmental metrics. Some firms are developing new metrics, creating tools to prioritize these indicators, and experimenting with more qualitative issues of human health and ecosystem impacts (Strang and Sage, forthcoming; Wright et al., 1998). A few have even begun to investigate the broader social dimensions of industrial activity (Davis, 1998).

As can be seen in Table E-1, most metrics focus on environmental burdens and ecoefficiency concerns such as resource use, water pollution, air emissions, and waste disposal. Far fewer metrics link these burdens to actual impacts (e.g., increased incidence of disease, declining health of a water body). Measuring impacts, industrial or otherwise, presumes that one can reliably identify causal links with specific activities. Current tools to perform such analyses are crude. Ecosystem impacts, human health effects, habitat loss, and global climate change represent concerns for which new metrics are required.

The need for improved metrics will become even more important as greater attention is given to the concept of sustainability, a term generally accepted as indicating movement beyond ecoefficiency to describe "development that meets the needs of the present without compromising the ability of future generations to meet their own needs" (World Commission on Environment and Development, 1987). Incorporating sustainability will require a much better understanding of the synergy of environmental effects and the impact of varying scale (both temporal and spatial) across ecosystems.

In addition to identifying some of the technical shortcomings of current metrics, this report also underscores the dual role of environmental metrics as a tool to facilitate effective management and a critical element of public accountability. Many regulations impose a reporting burden on firms, and these reports require the use of certain metrics. Today, the value of environmental performance information is under threat because of the great number of different approaches for collecting and using such data (White and Zinkl, 1997; World

Business Council for Sustainable Development, 1998). Because of this diversity, it is often difficult to develop sufficiently comparable information on environmental performance across a single company, let alone a whole sector or nation. This problem is compounded by international differences. Indeed, one of the most important challenges is to devise metrics that serve the specific needs of users while simultaneously contributing to greater comparability across firms, industries, and nations. There has been some movement toward more standardized metrics (much of this facilitated by government requirements), but lack of comparability among companies is still a significant barrier to progress (White and Zinkl, 1997).

Lack of standardized metrics may also have contributed, at least in part, to the slow diffusion of best practices to small and medium-sized companies. At present, larger corporations are much more likely to investigate, develop, and use measures of environmental performance (Ehrenfeld and Howard, 1996; KPMG Denmark, 1997b). While larger companies have been on the forefront of implementing environmental metrics, the areas of the product life cycle under their direct influence often contribute only a fraction of the overall environmental impact imposed by their products. Some companies are now beginning to think about the degree of influence they can reasonably exert over life-cycle elements controlled by others, such as supply-chain firms, product users, and those concerned with end-of-life product disposition (Brown, 1998; Graedel and Allenby, 1998; Institute of Electrical and Electronics Engineers, 1997). The expansion of producer accountability over more of the product life cycle amounts to a first step beyond ecoefficiency toward more sustainable practices.

Finally, as we gain a better understanding of the environmental performance of industry, attention is turning to other contributors to environmental "load." Public-sector enterprises such as federal facilities, while creating significant environmental burdens, are only now beginning to come under the same scrutiny to which industry has become accustomed (United States Congress, Office of Technology Assessment, 1989; United States Environmental Protection Agency, 1983). Large municipalities also are responsible for significant environmental impacts. Finally, it should be noted that industrial environmental metrics are now beginning to expand beyond manufacturing to the service sector, including such disparate industries as retail sales, distribution services, airlines, energy services, and health care (Chertow and Esty, 1997; Environmental Law Institute, 1998; Graedel, 1997). Presently, the lack of service-sector metrics makes it difficult even to begin to quantify the impacts of these businesses or their potential for environmental improvement.

In summary, the committee has identified several challenges to the pursuit of a broader, more robust, and effective system of environmental metrics:

- improving standardization of metrics both within and across industry sectors;

- disseminating best practices more widely;
- applying metrics across the entire supply chain and product life cycle;
- developing new analytic tools (e.g., impacts vs. loads); and
- addressing emerging environmental issues (e.g., ecosystem health, sustainability).

## A FRAMEWORK FOR ACTION

Over the past several decades, public concerns about risks to human and ecosystem health have driven individuals and organizations to act in a more environmentally conscious manner (Council on Environmental Quality, 1995; United States Environmental Protection Agency, 1992, 1996). The committee believes this trend will likely continue as scientific understanding of environmental systems improves and society's demands for environmental improvement persist. As public attitudes continue to "raise the bar" with respect to environmental performance, each economic sector (e.g., agriculture, industry, municipalities) will choose the methods most suited to meeting these challenges. If a particular sector's performance in the environmental arena is seen as inadequate and if social pressure is maintained, actions for meeting expectations will likely be prescribed. In the case of industry, past experiences with this prescriptive process have been viewed as particularly intrusive and inefficient. Limiting the imposition of new regulations will require skillfully harnessing performance improvements that have environmental or economic benefit while also formulating innovative strategies to efficiently address issues that lack financial incentives. The committee is convinced that environmental performance metrics will play an important role in these efforts, providing a valuable tool to industry as it strives to do its part to reduce the impact of human activities on the global environment.

The committee observes that concerns over compliance have driven the majority of environmental performance improvements to date. More recently, the private sector has discovered there are real rewards for taking a more proactive approach to protecting the environment. Individual companies and industry associations are becoming increasingly interested in, and capable of, contributing solutions to environmental challenges. As the private sector continues to demonstrate a greater capacity to drive environmental improvement, the government's role should begin to shift from that of a regulator to that of a facilitator. In the future, partnerships between government, industry, and citizen groups will likely yield more creative and efficient solutions to environmental problems. This is not to say that government should abdicate its leadership on environmental issues. Environmental quality is a "public good," and for this reason a significant, if perhaps declining, government role should be maintained. While acknowledging this, the committee emphasizes that better results and greater efficiency have generally been obtained from those companies that have voluntarily undertaken serious attempts at environmental improvement (Buzzelli, 1996).

## ESTABLISHING A BASELINE: BEST PRACTICES

Analysis of the metrics in use by the more progressive organizations engaged in this study yielded a number of similarities in terms of the types of measurements tracked (Table E-1). The committee feels that these environmental performance metrics, in their entirety, represent a broadly accepted set of "best practices." Best practices, however, are far from common. Therefore, some guidance should be provided to those organizations that have yet to establish a comprehensive framework of environmental metrics.

*RECOMMENDATION 1:* **Companies should investigate and implement to the greatest degree practicable environmental metrics representative of current best practices. Based on the four sector studies and the experience of its members, the committee urges firms to develop metrics in the 15 categories described in Table E-2.**

TABLE E-2   Recommended Categories for Environmental Performance Metrics in Manufacturing and Product Use

| Category | Brief Description and Examples |
| --- | --- |
| *Manufacturing Related* | |
| Pollutant releases | Includes:   *Air*—Some data collected to meet regulatory reporting requirements (i.e., Toxic Release Inventory [TRI]).  Separated into hazardous/nonhazardous.[a] <br> *Water*—Some data collected to meet reporting requirements.  Similar to above. <br> *Solid*—Some data collected to meet reporting requirements.  Similar to above. |
| Materials use/efficiency | Separated into hazardous/nonhazardous.[a] |
| Energy use/efficiency | Broken down by resource (e.g., petroleum, natural gas, coal, renewable).  Some companies have also begun to assess in terms of global warming potential (e.g., $CO_2$ equivalents). |
| Water use/efficiency | May track process water and cooling water separately. |
| Greenhouse gas emissions | Separated by gas (e.g., $CO_2$, $CH_4$, $N_2O$).  Can be expressed in $CO_2$-warming-potential equivalents. |
| Percent reuse/recycle/ disposal | Useful for assessing the use of individual process inputs as well as the final disposition of some intermediate products. |
| Packaging | Measured on either an absolute or per-product basis. |
| Land use | Separated into percent of land preserved, land developed, land restored, and inactive or abandoned developed land. |
| Environmental incidents | Classified by regulatory violations, fines, permit exceedances, accidents, etc. |
| Health and safety | Incidence of employee illness and injury and hours of training taken in safety, hazardous waste handling, etc. |

*continued*

TABLE E-2  *Continued*

| Category | Brief Description and Examples |
|---|---|
| *Product-Use Related* | |
| Pollutant releases | Includes:  *Air*—Separated into hazardous/nonhazardous[a] (e.g., emission standards for automobiles).<br>*Water*—Same as above (e.g., output water quality of a washing machine or dishwasher).<br>*Solid*—Same as above (e.g., toner cartridge for printer or copier). |
| Materials use/efficiency | Materials required for product use (e.g., detergent in cleaning appliances, fluids in automobiles). |
| Energy Use/efficiency | Energy requirements for product use (e.g., corporate average fuel economy [CAFE] in auto industry, power use in electronic devices [U.S. Environmental Protection Agency's (EPA) Green Lights program], cooking efficiency [Electrolux]). |
| Water use/efficiency | Water requirements for product use (e.g., appliances, toilets). |
| Greenhouse gas emissions | Primarily a function of energy use.  Can be expressed in terms of $CO_2$ (or $CO_2$ equivalents). |
| End-of-life disposition | Units or amounts of product reused, recycled, or disposed of (may be further separated by method of disposal). |

[a]All references to a hazardous/nonhazardous distinction are made with respect to existing regulatory definitions in the United States.
NOTE: In many cases the usefulness of metrics will be enhanced by appropriate normalization (e.g., per unit product, per unit sales, per product use, per product lifetime).

The categories listed in Table E-2 fulfill several requirements.  First, reliable and relatively unambiguous measurements may be derived in each of these categories based on present knowledge and technology.  Second, many of these categories relate directly to core business concerns (e.g., materials use, energy efficiency), regulatory requirements, or the maintenance of good relationships with the community or with regulators.  Lastly, these metrics require information that is already collected by most companies as a matter of compliance, inventory, or waste management.  That is not to say that resources will not be required to assemble the information into a usable form, but the means for obtaining many of these data is presently in place.

While the relevance of individual categories may vary by industry (the paper industry will, for instance, have greater interest in land-use metrics than the electronics industry), most have broad applicability.  Recommendation 1 provides guidance to the many companies that have yet to undertake a comprehensive program of environmental performance measurement, as well as a check for those companies with programs already in place.  The committee feels that these categories collectively represent an instructive and broad assessment of the present state of an organization's environmental performance and one that may be reasonably achieved. Many further improvements to environmental performance measurement and reporting continue to be needed. However, the widespread implementation of these metrics will be a significant and meaningful first step.

## GOALS FOR IMPROVING
## INDUSTRIAL ENVIRONMENTAL PERFORMANCE

To assist industry efforts to improve environmental performance, the committee has identified five goals for enhancing the development and use of industrial environmental performance metrics. Each goal is accompanied by one or more specific recommendations.

### Goal 1: Adopt Quantitative Environmental Goals

The establishment of quantitative goals provides a clear point of reference for developing and applying metrics. This is true at both the national and the corporate levels. Unless management and employees have a firm grasp of exactly what is expected of them and the criteria by which they will be evaluated, attention to environmental issues will be unfocused and motivation will wane. At the national level, setting quantitative goals has a dual benefit. First, such goals provide policy makers and government officials with clearly articulated objectives on which to focus throughout the sometimes convoluted political process. Second, industry benefits from a greater degree of certainty, especially when assessing regulatory expectations within the context of future planning activities.

*RECOMMENDATION 2*: **The U.S. government should strengthen its role in setting and reporting progress toward national environmental goals.**

The federal government has a singular role to play in bringing together the technical expertise to prioritize the myriad environmental issues of national concern and to periodically update these assessments. Previous exercises of this type have been attempted (e.g., President's Council on Sustainable Development), but rarely have they articulated an explicit ranking of environmental priorities or established quantitative benchmarks.

*RECOMMENDATION 3:* **Individual companies and industry sectors should set quantitative environmental goals and track and report their progress in meeting these goals. Individual companies and industry sectors should take the initiative in setting, tracking, and reporting on their progress in meeting quantitative environmental performance goals.**

Companies that set quantitative environmental goals and commit to tracking and reporting progress often realize rapid improvements in environmental performance. While national environmental goals may not always be directly applicable, related measures should be incorporated into corporate planning to the degree practicable. The committee recognizes that important issues associated with ecosystem health, biodiversity, and sustainability still largely defy attempts at quantification, although progress continues to be made in this area.

### Goal 2: Improve Methods of Ranking and Prioritizing Environmental Impacts

Efforts must be undertaken to develop an acceptable system for prioritizing the issues of greatest environmental concern. Doing this will require moving from the measurement of environmental loads (e.g., air emissions, water emissions, resource use, land use) to the measurement of environmental impacts (e.g., human health impacts, ecosystem impacts). With such a system, goals and metrics can be established so that scarce public and private resources are directed toward reduction of environmental impacts in the most effective manner.

Companies often invest in a variety of voluntary environmental initiatives. Some of these efforts result in cost-effective lowering of environmental impacts and some do not. Such a framework would be valuable to industry and government as they continually seek to reassess and update their environmental goals. (See Recommendations 2 and 3.)

*RECOMMENDATION 4:* **Develop categorization systems to prioritize and target opportunities for reducing environmental impact. The U.S. government should facilitate a process with academia, industry, state agencies, and nongovernmental organizations to develop improved methods of ranking, categorizing, and prioritizing the relative impact of industrial environmental loads.**

This process should begin by focusing on human health risks and extend to issues of ecosystem health and long-term sustainability as knowledge and understanding of environmental systems evolve. Present knowledge may not allow for explicit numeric scoring of all impacts under all circumstances, but the committee feels that sufficient knowledge does exist to begin to prioritize categories of environmental loads (e.g., air emissions, water emissions, resource use, land use) relative to one another. The need for prioritization applies both within and across respective categories. Within the category of hazardous emissions, a number of efforts have been undertaken to rank substances with respect to their impacts on human health. Emissions may, therefore, provide a useful starting point. Data collected under TRI, as well as hazardous waste generation and disposal data collected under the Resource Conservation and Recovery Act, may not fully represent a facility's environmental impact, but they do provide two of the only examples of consistent cross-industry metrics. Methodologies must also be designed to compare the relative environmental impacts of hazardous emissions against other categories of environmental loads, such as materials use, land use, or waste disposal.

The committee believes that the absence of ranking systems lessens the value of environmental reporting. Because many of these topics are beyond the province of industrial research, government should assume a leadership role in bringing together industry leaders, academics, and public stakeholders to investigate metrics and goals that reflect cumulative and long-term environmental impacts.

### Goal 3: Improve the Comparability or Standardization of Metrics

Corporate managers, local communities, and a host of other stakeholders desire greater uniformity in environmental metrics. Comparable metrics support better internal decision making; allow for benchmarking across industries; and provide much wanted information to investors, lenders, citizen groups, and others. Reporting standards will also help make credible to the public industry efforts to improve environmental performance.

*RECOMMENDATION 5:* **The U.S. government should facilitate a process of establishing consistent, standardized industrial environmental metrics through the involvement of experts from industry, nongovernmental organizations, and federal agencies.**

The absence of standard metrics impedes benchmarking and reduces the value of public reporting of environmental performance. While reporting of standardized metrics is voluntary, a company claiming to have reduced its environmental burden should be judged by objective criteria (not unlike the formal definitions of terms like "low fat" or "nonfat" recently instituted in the food-processing industry). It is also possible that peer pressure may begin to push companies to report environmental data as more of their competitors choose to do so.

Developing a set of standard metrics is absolutely critical to establishing a pattern of continual improvement in industrial environmental stewardship. While government participation is prudent in any process that seeks to set national standards, industry should play an integral role in the development, implementation, and promotion of standardized environmental performance metrics. At the level of the individual facility or firm, senior business leaders need to work with environmental managers to encourage the adoption of, at a minimum, measures of ecoefficiency and toxic dispersion. The committee stresses that a company's selection of which metrics to use and report should remain voluntary, but organizations claiming improved environmental performance need to make their case on the basis of accepted standards.

*RECOMMENDATION 6:* **Promote standardized industrial environmental performance metrics in international forums.**

The world is becoming increasingly interconnected, and the environmental performance of a U.S. corporation is often judged by more than local or national standards. The same applies to foreign multinationals operating in the United States. Some method of comparing environmental performance across countries is required. The U.S. government can play an important role in promoting standardized metrics internationally.

In today's global economy, corporate operations are not limited by national boundaries but depend on extensive global supplier chains and distribution networks. Standardized metrics, should they become globally accepted, will help level the playing field with respect to the environmental practices of companies operating under very different regulatory systems. Although establishing international industrial environmental performance standards will not be easy, efforts should be made to bring global attention to this issue.

## Goal 4: Expand the Development and Use of Metrics

The time is right to expand the use of environmental performance metrics over more of the product life cycle and to disseminate knowledge of best practices to a wider audience. In recent years some of the largest manufacturers have been providing more-detailed quantitative information on the environmental dimensions of their operations. Environmental measures in the manufacturing stage are important, but attention must now shift to other life-cycle areas. One major challenge is encouraging the development of metrics within the manufacturer's supply chain. The committee recognizes there are limits to how deeply life-cycle attributes (and the supply chain) can reasonably be investigated. Still, it believes that the potential environmental benefits of viewing the product life cycle more holistically demand that the corporate boundaries of environmental performance metrics be enlarged.

*RECOMMENDATION 7:* **Industry should integrate environmental performance metrics more fully throughout the product life cycle.**

Few companies or industries control their product from cradle to grave, and many exercise direct influence over only a fraction of product life cycles. Industrial executives, managers, and engineers should begin to extend the application of environmental performance metrics both up (e.g., to account for product use and end of life) and down (e.g., to account for raw materials acquisition and processing) the supply chain.

The committee recommends that companies and industrial sectors take the lead in more fully assessing the life-cycle impacts of their products. This process may include conducting surveys or studies in partnership with suppliers to determine which metrics are most useful and feasible. The committee believes that an invaluable contribution to this effort would be the establishment of a consistent, cost-effective methodology for approximating relative life-cycle impacts. A systems approach will be required, if suppliers, manufacturers, consumers, and those responsible for the final disposition of a product can reasonably assess their roles in lessening the overall environmental impact of their activities.

Small and medium-sized enterprises could benefit from greater exposure to the more advanced environmental practices of larger corporations, as could other large firms that have been somewhat less progressive. While companies cannot be expected to release proprietary information or methods, greater efforts must be made to transfer to other levels of industry the corporate practices more common to large manufacturers.

*RECOMMENDATION 8:* **The U.S. government, acting in concert with industry, should gather and disseminate information on best practices in industrial environmental performance measurement. Improved efforts must be made to transfer the knowledge and technology of these methods across industries and sectors, particularly to small and medium-sized enterprises.**

Approaches to measuring and improving environmental performance are proliferating throughout the world. Some system needs to be devised that more effectively communicates these techniques to small and medium-sized companies, as well as to larger companies that have yet to develop environmental measures. The Internet allows for the creation of a widely accessible clearinghouse of environmental metrics information. An appropriate government agency (e.g., EPA, U.S. Department of Commerce) should engage expertise from the private and public sectors to assemble and periodically update an online library of industry-specific metrics and case studies. Other avenues of dissemination might include state agencies and industry associations.

### Goal 5: Develop Metrics that Keep Pace with New Understanding of Sustainability

Society's understanding of and commitment to the concept of sustainability is increasing. As this interest grows, all sectors of the economy must begin to investigate methods of assessing and improving the sustainability of their activities. While many companies have made headway in the application of ecoefficiency metrics and programs, this represent only a first step toward sustainability.

> *RECOMMENDATION 9:* **The U.S. government and industry should assure that adequate research attention is directed toward furthering understanding of the complex environmental interactions associated with sustainability.**

Industry's role will be driven primarily by competitive pressures as customers, investors, and regulators demand better environmental attributes from products and processes. Government's role will be to examine ways in which industrial operations and products affect the various aspects of sustainability. This may involve investigating the implications of long-term industrial activity on the environment, including such issues as materials flows and energy use.

While the concept of sustainable development has widespread appeal, there is as yet no scientific consensus on a definition of the concept or indices by which it may be measured at the macro, or societal, level. Assessing a given industrial product, process, technology, or facility with regard to sustainability will require the development of systems approaches for which very few relationships have yet been developed. In addition, although attention to purely environmental issues is important, it should be noted that economic and social concerns are integral to the concept of sustainability.

> *RECOMMENDATION 10:* **Conduct research on methods of integrating socioeconomic criteria into sustainability measures.**

Research is needed to help solve the analytic, relational, and informational challenges associated with sustainability. These challenges involve not only single-issue complexities (e.g., related to the environment) but also those relating multiple behaviors and impacts (e.g., economic-environmental, social-environmental) and varying scales (e.g., local, regional, global).

Presently, it is difficult to directly relate a firm's contribution to many broad measures of economic and social status (e.g., per-capita income, average education level). Researchers must begin to examine methods of analysis and metrics that address society's ability to link environmental, economic, and social activities in a manner that can guide progress toward sustainability. These methods and metrics often reach far beyond industry's sphere of influence. Nonetheless, as a strong force within society, industry must begin to investigate its role in this process. The committee notes that current concepts and understanding of sustainability are incomplete.

## CONCLUSIONS

With new knowledge and changing public expectations, fresh environmental challenges are arising that are not addressed by existing environmental metrics. Ecosystem impacts, human health effects, habitat loss, and global climate change are among a few of the emerging issues for which metrics are needed. To realize the full potential of environmental metrics will require changes in industry, in government, and at the community level. All of these sectors have important roles to play, not only in improving industrial practices but also in extending the lessons learned there to the vast array of other societal activities that impact the environment. Much work remains, but as society moves to achieve sustainable development, environmental metrics will provide a valuable tool for influencing decision making and driving innovation.

## REFERENCES

Brown, M. 1998. Working with suppliers to improve environmental performance. Paper presented at the National Academy of Engineering International Conference on Industrial Environmental Performance Metrics, November 1–4, 1998, Irvine, Calif.

Buzzelli, D.T. 1996. The Next Industrial Frontier: Managing the Business of Environment in Asia and the Pacific. Presentation at the Asia Pacific Responsible Care Conference, Beijing, China, September 18–20, 1996.

Chertow, M.R., and D.C. Esty. 1997. Environmental policy: The next generation. Issues in Science and Technology 14:1.

Council on Environmental Quality. 1995. 25th Anniversary Report of the Council on Environmental Quality (1994-1995). Available online at http://ceq.eh.doe.gov/reports/1994-95/rep_toc.htm. [February 12, 1999]

Davis, J. 1998. Exploring sustainable development: The World Business Council on Sustainable Development Global Scenarios. Paper presented at the National Academy of Engineering International Conference on Industrial Environmental Performance Metrics, November 1–4, 1998, Irvine, Calif.

Ditz, D., and J. Ranganathan. 1997. Measuring Up: Toward a Common Framework for Tracking Corporate Environmental Performance. Washington, D.C.: World Resources Institute.

Ditz, D., J. Ranganathan, and R.D. Banks. 1995. Green Ledgers: Case Studies in Corporate Environmental Accounting. Washington, D.C.: World Resources Institute.

Ehrenfeld, J., and J. Howard. 1996. Setting environmental goals: The view from industry. Pp. 281–325 in Linking Science and Technology to Society's Environmental Goals. Washington, D.C.: National Academy Press.

Environmental Law Institute. 1998. Proceedings of the symposium, Impacts and Leverage: Environmental Management and the Service Sector, March 31, 1998, Washington, D.C.

Epstein, M.J. 1996. Measuring Corporate Environmental Performance: Best Practices for Costing and Managing an Effective Environmental Strategy. Chicago: Irwin Professional Publishing.

Fischer, K., and J. Schot. 1993. Environmental Strategies for Industries—International Perspectives on Research Needs and Policy Implications. Washington, D.C.: Island Press.

Graedel, T.E. 1997. Life-cycle assessment in the service industries. Journal of Industrial Ecology 1(4):57–70.

Graedel, T.E., and B.R. Allenby. 1998. Industrial Ecology and the Automobile. Upper Saddle River, N.J.: Prentice Hall.

Institute of Electrical and Electronics Engineers (IEEE). 1997. Proceedings of the 1997 IEEE International Symposium on Electronics and the Environment, May 5–7, 1997, San Francisco.

KPMG. 1997. International Survey of Environmental Reporting. Lund, Sweden: KPMG.

KPMG Denmark. 1997a. Environmental Reporting. Copenhagen: Silkeborg Bogtryk.

KPMG Denmark. 1997b. The Environmental Challenge and Small and Medium-Sized Enterprises in Europe. The Hague: KPMG Environmental Consulting.

Popoff, F. 1993. Full-cost accounting. Chemical and Engineering News 71(2):8–10.

Porter, M.E. 1991. America's green strategy. Scientific American 264:4.

Strang, R., and L. Sage. Forthcoming. Measuring environmental performance through comprehensive river studies. In Measures of Environmental Performance and Ecosystem Conditions, Peter Schulze, ed. Washington, D.C.: National Academy Press.

3M. 1998. Reducing Waste and Environmental Releases. Available online at http://www.mmm.com/profile/envt/epr/releases.htm. [February 12, 1999]

United States Bureau of the Census. 1997. Statistical Abstract of the United States. Washington, D.C.: U.S. Government Printing Office.

United States Congress, Office of Technology Assessment. 1989. Facing America's Trash: What Next for Municipal Solid Waste? OTA-0-424. Washington, D.C.: U.S. Government Printing Office.

United States Environmental Protection Agency (USEPA). 1983. Final Report of the Nationwide Urban Runoff Program. PB84-185545. Washington, D.C.: USEPA Water Planning Division.

United States Environmental Protection Agency (USEPA). 1992. National Air Quality and Emissions Trends. EPA-450-R-92-001. Research Triangle Park, N.C.: USEPA Office of Air Quality Planning and Standards.

United States Environmental Protection Agency (USEPA). 1995a. Environmental Accounting Case Studies: Full Cost Accounting for Decision Making at Ontario Hydro. EPA742-R-95-004. Washington, D.C.: USEPA Office of Pollution Prevention.

United States Environmental Protection Agency. 1995b. Environmental Accounting Case Studies: Green Accounting at AT&T. EPA742-R-95-003. Washington, D.C.: USEPA Office of Pollution Prevention.

United States Environmental Protection Agency. 1996. The U.S. EPA's 25th Anniversary Report: 1970–1995. Available online at http://www.epa.gov/25year/. [May 15, 1999]

White, A., and D. Zinkl. 1996. Corporate Environmental Performance Indicators: A Benchmark Survey of Business Decision Makers. Boston: Tellus Institute.

White, A., and D. Zinkl. 1997. Green Metrics: A Status Report on Standardized Corporate Environmental Reporting. Boston: Tellus Institute.

World Business Council for Sustainable Development (WBCSD). 1998. Eco-efficiency Metrics and Reporting: State-of-Play Report. Geneva: WBCSD.

World Commission on Environment and Development. 1987. Our Common Future. New York: Oxford University Press.

Wright, M., D. Allen, R. Clift, and H. Sas. 1998. Measuring corporate environmental performance: The ICI environmental burden system. Paper presented at the NAE Workshop on Industrial Environmental Performance Metrics, January 28–29, 1998, Washington, D.C.

# PART I

# Why and What

# 1

# Why Study Environmental Metrics?

## OVERVIEW AND OBJECTIVES

The ability to gauge improvement in any endeavor is critically dependent on establishing valid methods of measuring performance. Tracking progress toward an established goal serves to influence behavior by providing continual feedback, and it requires reliable and consistent metrics against which performance can be compared.

In response to competitive pressures and the increasing informational demands of a wide range of stakeholders (e.g., customers, lenders, investors, regulators, local communities), many companies have begun investigating metrics that will improve their ability to assess the environmental aspects of their operations. Growing interest in environmental metrics, however, has exposed the considerable lack of coordination that currently characterizes their development and use.

The measurement of industrial environmental performance is still in its infancy, but the practice has a high potential for growth, if industry's common use of sophisticated financial metrics is any indicator. Many of the financial measures developed in recent years would mystify Wall Street analysts from past eras, yet today they are seen as indispensable. As customers, investors, and regulators have demanded better financial information, companies have devised ever more detailed and instructive indicators to meet the demand. A similar evolution has now begun to take place in the area of industrial environmental performance metrics.

Measures of environmental performance abound, spanning a wide range of issues related to governmental, societal, and commercial behavior. The word "industrial" defines the scope of this study. Despite the attention industry receives, there are many environmental effects attributable to human activity in which industry plays little role. It can even be argued that within many developed nations (e.g., those within the Organization for Economic Cooperation and Development), significant improvements in industrial environmental performance over the past 25 years may have lowered the share of environmental impact attributable to industry. As such, consideration must be given to assessing which issues lie within, and outside, industry's sphere of influence.

Industry has a unique set of motivations and constraints with regard to the environment. Industry's primary goal is to service the needs and wants of its customers. In the process, the industrial sector often contributes products, services, and research that improve the quality of life for much of the world's population. The extent to which a company's products or services meet society's demands also determines how well the firm fulfills a primary corporate objective: maximization of shareholder value. The pursuit of increased shareholder value is, however, generally subject to societal and governmental constraints that inhibit practices viewed as socially detrimental. One of these is the imposition of an undue burden on the environment. As with other social checks, the environmental constraint is a product of governmental regulation, liability concerns, moral imperatives, popular sentiment, and customer demands. In the future, these factors are likely to increase industry's motivation to reduce environmental impact.

Industry has been advancing along an environmental management learning curve since the early 1970s (Figure 1-1). Twenty-five years ago, protecting the environment was viewed as a drag on economic progress, and there was considerable resistance. The mode was reactive. In the 1980s some companies began to see that proactive or anticipatory approaches could help lessen environmental burden. There was an emergence of corporate pollution prevention plans and, later, of more comprehensive environmental management systems. In the 1990s industry began to develop a more comprehensive approach to assessing environmental costs based on the emerging principles of full-cost accounting (Banks, 1995; Thayer, 1995). Industry's environmental goals and metrics have also evolved over the past quarter-century (Table 1-1).

Even among those skeptical of the it-pays-to-be-green paradigm, there is acknowledgment that all signs point toward continuing social pressure for improved levels of environmental performance (Dingell, 1990). As a result, companies capable of reacting swiftly and efficiently to the changing landscape will find themselves at a distinct advantage. Responding to, or in some cases anticipating, these changes requires not only the ability to vary one's products and processes to lessen environmental burdens but also to communicate improvement to a broad range of stakeholders. Successful companies in such a scenario will be

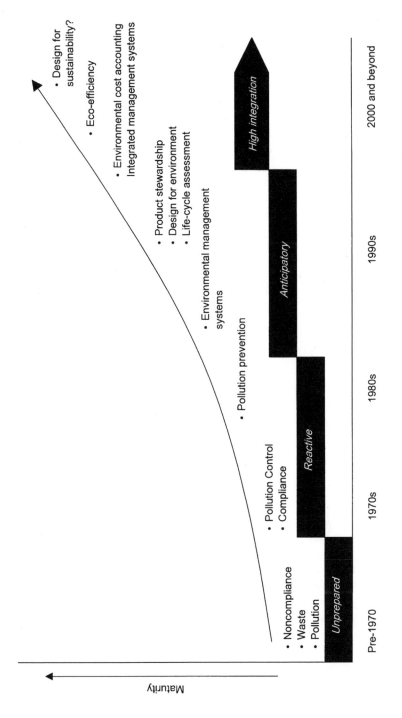

FIGURE 1-1  Industry's environmental design and management learning curve.  SOURCE:  Adapted from Richards and Frosch (1997).

TABLE 1-1    Evolution of Industry's Environmental Goals and Metrics

| Pre-1970 | Mid-1970s–Mid-1980s | Mid-1980s–Mid-1990s | 2000 and Beyond |
|---|---|---|---|
| *Goals* | | | |
| • None | • Meet regulatory standards | • "Cost" avoidance<br>• Emissions reduction<br>• Preempt regulations<br>• Image<br>• Leadership<br>• Legitimacy protection<br>• Competitive edge | • Explicit mainstreaming of environmental goals<br>- Design for environment<br>- Life-cycle assessment<br>- Environmental cost management<br>- Recycling targets<br>- Alternative products |
| *Metrics* | | | |
| • None | • As required by regulation<br>- hazardous waste generation<br>- permit violations<br>- payout of regulatory fines | • As required by regulation<br>• Rates and quantity of emissions<br>• Energy efficiency metrics<br>• Productivity improvement metrics<br>• Toxicity reduction metrics | • As required by regulation<br>• Rates and quantity of emissions<br>• Energy efficiency metrics<br>• Productivity improvement metrics<br>• Toxicity reduction metrics<br>• Recycle rates<br>• "Weighted" metrics |

those that find new approaches for meeting consumer demands while lowering environmental impact.

While many industries have significantly improved their environmental performance in recent decades, the committee observes that the majority of these improvements have been driven by government regulation. Consequently, metrics imposed by regulation have dominated. As industry begins to set environmental goals that move beyond compliance, new methods of measuring and tracking improvement need to be developed. To investigate potential advancement in the area of industrial environmental performance metrics, the National Academy of Engineering has undertaken this study with the following four objectives:

- examine U.S. industrial experience in setting goals and measuring progress in environmental performance;
- identify factors that have contributed, or will contribute, to improved industrial environmental performance in the United States;
- assess the relative successes and shortcomings of current methods of measuring industrial environmental performance; and
- recommend a set of industrial environmental performance metrics that define current best practices and identify directions for future improvement.

Metrics provide a means of identifying trends and assessing performance with respect to a goal. Many goals will have strictly defined endpoints, for example the concentration of a pollutant in wastewater effluent or an explicitly stated reduction (e.g., 50 percent) in emissions. More broadly stated objectives such as increasing sales revenue or lowering environmental impact have some usefulness, but the committee believes that without including quantitative assessment criteria the usefulness of such objectives for measuring and motivating progress is limited. While defining environmental goals for society, industry, or individual companies is not the purpose of this study, goals are often an important precondition to establishing robust metrics.

Many corporate goals flow, directly or indirectly, from national or state environmental objectives, generally communicated in the form of regulations. Compliance has been and will remain a powerful motivation, but increasing evidence suggests that improving environmental performance can also yield business-related benefits (Deutsch, 1998; Esty and Porter, 1998; Hart and Ahuja, 1996; Porter and van der Linde, 1995). Some of the recent interest in scrutinizing environmental performance has been driven by the growing realization that traditional methods of accounting often do not fully capture or identify environmental costs (Epstein, 1996; Ditz et al., 1995; United States Environmental Protection Agency, 1995). Corporate accounting procedures frequently place environmental expenditures into nebulous categories such as "overhead," thereby making it difficult to break out individual costs. Other problems can arise, as capital budgeting procedures frequently do not allow for the separate identification of environmental costs as part of investments in new equipment or processes.

As companies have improved their ability to relate environmental costs to the bottom line, management's interest in assessing environmental performance has grown (Rappaport and MacLean, 1995). Interest in metrics capable of tracking environmental performance has followed. The emergence of the new field of environmental, or full-cost, accounting is seen by many as the next step in helping companies move from reactive compliance-based performance toward more proactive strategies (Banks, 1995; Fischer and Schot, 1993; Popoff, 1993; Richards and Frosch, 1997).

Despite the assertion by a number of researchers and companies that financial rewards await organizations that improve their environmental performance, there are admittedly limits to this model. Realistically, there are always wastes associated with most industrial processes and often significant costs involved in reducing them. Some maintain that lessening environmental burdens, at least in response to compliance standards, inhibits economic progress (Walley and Whitehead, 1994). Another group of researchers finds little correlation, positive or negative, between the bottom line and improved environmental performance (Jaffe et al., 1995).

As should be apparent, no conclusive evidence or overarching theory has yet been proffered to conclusively link environmental performance and profitability. It is quite likely that the truth, as in most things, depends on individual circumstances (Clarke, 1994). Still, there is little dispute that when tailored to the needs of an individual firm or industry, environmental performance metrics have provided a means of improving both the environmental and financial performance of a growing number of companies.

As this paradigm takes hold, some companies have begun to view environmental performance as a means of advancing primary business objectives (Magretta, 1997; Van Epps and Walters, 1996). These firms have started to incorporate environmental information into decisions on such topics as product selection, marketing, and strategic planning (Fussler and James, 1996; Hart, 1997). Following the adage that what gets measured gets managed, these companies have started down the path to "ecoefficiency," a term the World Business Council for Sustainable Development (1998) defines as

> ". . . the delivery of competitively priced goods . . . while progressively reducing ecological impact and resource intensity . . . to a level at least in line with the earth's estimated carrying capacity."

While some companies and industrial associations have begun to research, test, and implement ecoefficiency metrics, little consensus has emerged as to which are most useful.

Beyond direct financial incentives, the committee also notes that the development and use of improved metrics have been encouraged by the public's growing awareness of environmental issues. Rising interest in environmental performance has led to increasing attention being given to those things that, in a larger societal context, "should" be measured. The environmental load imposed by resource use, water pollution, air emissions, waste disposal, and other consequences of a highly developed economy have prompted questions regarding how actual impacts might be lessened. As a direct and indirect contributor to the global environmental burden, industry has begun to investigate the extent of its role and potential mitigating actions. Detailed exploration by the committee of how environmental impacts might be measured and characterized reveals that the process will not always be straightforward.

Reducing environmental impacts presumes that one can reliably make a causal link between specific industrial activities and specific impacts. This is an area where considerable uncertainty still exist (National Research Council, 1997). While headway has been made in establishing links between some toxic substances and human health (United States Environmental Protection Agency, 1998a, b), connections between many social, industrial, and commercial activities and ecosystem health are less well defined. The need for clear understanding will become even more important as increasing attention is given to the concept of sustainability. Though a somewhat ambiguous term, sustainability is generally accepted as indicating movement beyond ecoefficiency. (See Chapter 11.) The World Commission on Environment and Development (1987) describes sustainable development as "development that meets the needs of the present without compromising the ability of future generations to meet their own needs."

Incorporating sustainability concerns into industrial management will require a much better understanding of how synergism and differences in temporal and spatial scale play out in complex environmental systems. It will also require an improved capacity to assess the regenerative and assimilative capacity of natural systems. Uncertainty and the high degree of interconnectedness of natural systems will make the task of identifying basic indicators difficult and, in some instances, possibly impractical. The committee believes a further challenge will be to devise a range of metrics capable of measuring all these complex interactions while still retaining a sufficient degree of simplicity and transparency. If such measures are not comprehensible, they will not be broadly useful.

Fulfilling all of these requirements is a formidable task. The committee feels that the ideal metric, or suite of metrics, will need to have distinct and clear meaning as it points the way toward the goals of increased industrial efficiency and profitability and environmental sustainability. It is the ambitious intent of this report to provide some insight into how an industry might measure progress toward evolving environmental goals (both society's and its own) through the development and use of improved environmental performance metrics.

The measurement of industrial environmental performance is stimulating considerable global interest. A number of efforts under way nationally and internationally are examining the issue. This effort differs from others in two key regards. First, it is an industry-centered study involving a committee primarily composed of professionals with corporate and manufacturing experience. As a result, the report may address environmental performance issues from a perspective that is different than that of many other studies. Second, in an attempt to provide more in-depth analysis rather than general commentary, this effort focuses on only four major manufacturing industries (automotive, chemical, electronics, and pulp and paper).

While a review of many sectors, such as the service industry, is notably absent in this report, the sectors the committee did examine in combination represent a significant portion of U.S. industrial capacity (12.6 percent of gross domestic product in 1995 [Bureau of the Census, 1997]). In addition to their size, these four sectors were chosen because of the diverse nature of their products and processes. Operations within the selected industries combine to span the full spectrum of product life cycle, as well as a wide range of market niches (e.g., raw materials, component parts, finished consumer products). These sectors may therefore be taken as somewhat representative of large-scale U.S. manufacturing, perhaps allowing for the extrapolation of at least some results across industries or nationally. Because another objective of this project was the dissemination of U.S. best practices to industries within countries of the Asia-Pacific Economic Cooperation, the relevance of the four sectors to the Pacific Rim economies also had some bearing on which were chosen.

Chapter 2 provides an introduction to industrial environmental performance metrics and the decisions they are intended to support. Also included is a discussion of the individual metric characteristics that companies, as well as commercial and public stakeholders, have found most useful. Chapters 3–7 contain an introduction and detailed case studies describing the past and present experiences of the four targeted industries. Chapters 8–10 review U.S. experiences while identifying present trends and challenges to improvement. Chapter 11 explores some of the long-term issues facing industry as it prepares to address expected future demands from the public for better environmental information. In this analysis, emphasis is on describing environmental performance metrics in terms of their links to related corporate goals identified by the committee. There is some discussion of how these goals are changing. Finally, Chapter 12 contains the committee's conclusions and recommendations.

## REFERENCES

Banks, R.D. 1995. Green costs and benefits. New York Times. July 9, Science Section.

Bureau of the Census. 1997. Statistical Abstract of the United States. Washington, D.C.: U.S. Government Printing Office.

Clarke, R.A. 1994. The challenge of going green. Harvard Business Review 72(4):37–50.

Deutsch, C.H. 1998. For Wall Street, increasing evidence that green begets green. New York Times. July 19, Business Section, p. 7.

Dingell, J.D. 1990. The environment and the economy: Striking a delicate balance. Pp. 137–143 in Environmental Policy and the Cost of Capital. Washington, D.C.: American Council for Capital Formation, Center for Policy Research.

Ditz, D., J. Ranganathan, and R.D. Banks. 1995. Green Ledgers: Case Studies in Corporate Environmental Accounting. Washington, D.C.: World Resources Institute.

Epstein, M.J. 1996. Measuring Corporate Environmental Performance: Best Practices for Costing and Managing an Effective Environmental Strategy. Chicago: Irwin Professional Publishing.

Esty, D.C., and M.E. Porter. 1998. Industrial ecology and competitiveness: Strategic implications for the firm. Journal of Industrial Ecology 2(1):35–43.

Fischer, K., and J. Schot, eds. 1993. Environmental Strategies for Industries: International Perspectives on Research Needs and Policy Implications. Washington, D.C.: Island Press.

Fussler, C., and P. James. 1996. Driving Eco-innovation: A Breakthrough Discipline for Innovation and Sustainability. Washington, D.C.: Pitman Publishing.

Hart, S.L. 1997. Beyond greening: Strategies for a sustainable world. Harvard Business Review 75(1):66–77.

Hart, S.L., and G. Ahuja. 1996. Does it pay to be green? An empirical examination of the relationship between emission reduction and firm performance. Business Strategy and the Environment 5:30–38.

Jaffe, A.B., S.R. Peterson, P.R. Portney, and R.N. Stavins. 1995. Environmental regulation and the competitiveness of U.S. manufacturing: What does the evidence tell us? Journal of Economic Literature 33:132–163.

Magretta, J. 1997. Growth through global sustainability: An interview with CEO Robert B. Shapiro. Harvard Business Review 75(1):78–83.

National Research Council. 1997. Building a Foundation for Sound Environmental Decisions. Washington, D.C.: National Academy Press.

Popoff, F. 1993. Full-cost accounting. Chemical and Engineering News 71(2):8–10.

Porter, M.E., and C. van der Linde. 1995. Green and competitive: Ending the stalemate. Harvard Business Review 73(5):120.

Rappaport, A., and R. MacLean. 1995. Greening the CFO: Recent practice and emerging trends. Corporate Environmental Strategies 2(4):2–9.

Richards, D.J., and R.A. Frosch, eds. 1997. The Industrial Green Game: Implications for Environmental Design and Management. Washington, D.C.: National Academy Press.

Thayer, A. 1995. Full accounting for environmental cost offers benefits to companies. Chemical and Engineering News 73(28):10–11.

United States Environmental Protection Agency. 1995. Environmental Cost Accounting for Capital Budgeting: A Benchmark Survey of Management Accountants. EPA742-R-95-005. Washington, D.C.: Office of Pollution Prevention and Toxics.

United States Environmental Protection Agency. 1998a. Ecotox Threshold. Available online at: http://www.epa.gov/oerrpage/superfnd/web/ oerr/r19/ecotox/index.html. [February 12, 1999]

United States Environmental Protection Agency. 1998b. Superfund: Introduction to the Hazard Ranking System. Available online at: http://www.epa.gov/oerrpage/superfnd/web/ oerr/ino_pro/ npl/hrs/hrsint.htm. [February 12, 1999]

Van Epps, R.E., and S.D. Walters. 1996. Measure for measure: Evaluating environmental performance with tact and insight. Corporate Environmental Strategy 3(2):42–48.

Walley, N., and B. Whitehead. 1994. It's not easy being green. Harvard Business Review 72(3):46–50.

World Business Council for Sustainable Development (WBCSD). 1998. Eco-efficiency Metrics and Reporting: State-of-Play Report. Geneva: WBCSD.

World Commission on Environment and Development. 1987. Our Common Future. New York: Oxford University Press.

# 2

# What Are Industrial Environmental Performance Metrics?

To help the reader understand the current use of environmental metrics within the four industries examined in this study, Chapter 2 describes a general metrics taxonomy, briefly reviews the characteristics that current literature and the committee's experience suggest are desirable in the formulation of useful metrics, and reviews the types of corporate decisions that are affected by environmental performance information. The chapter provides a general discussion only; more specific information and examples can be found in Appendixes A, B, and C.

## GENERAL CATEGORIES OF METRICS

Metrics developed for internal corporate use generally provide information on operations or management issues, but many also provide information useful to external stakeholders. Energy use, expressed either in absolute terms or on a per-unit-of-product basis, is an example of a measure of interest to both corporate managers and external stakeholders. The latter group (e.g., customers, regulators, investors, environmental groups) generally is interested in many internal corporate metrics and is concerned about the impacts of industry activities on the environment at the local, regional, and global levels. These concerns frequently relate to such issues as air quality, water quality, product recyclability, and regulatory compliance.

One widely used scheme of classifying metrics has been developed as part of the International Organization for Standardization (ISO) 14031 process (International Organization for Standardization, Annexes Testing Committee, 1996). This system groups metrics according to their utility in three areas:

- making operational decisions,
- making management decisions, and
- assessing the condition of the (external) environment.

## Operational Metrics

Operational metrics generally measure potential environmental burden in terms of inputs and outputs of materials and energy (Box 2-1). Operational metrics can sometimes be thought of as describing a rough "mass balance" for an industrial activity. Ditz and Ranganathan (1997) developed a classification system that separates operational metrics into four subcategories.

- Materials Use—Quantities and types of materials used (useful for tracking resource inputs and distinguishing their composition and source).
- Energy Consumption—Quantities and types of energy used or generated (provides the analog to materials use; made even more useful when fuel types are differentiated).
- Nonproduct Output—Quantities and types of waste created before recycling, treatment, or disposal (most useful in distinguishing production efficiency from end-of-pipe control solutions).
- Pollutant Releases—Quantities and types of pollutants released to air, water, and land (can be differentiated according to whether hazardous or nonhazardous or of global or local concern).

In many cases the comparability of each of these measures can be greatly enhanced by normalizing them with respect to such parameters as number of products produced or sales revenue. Efforts have also been made to normalize according to a product's value to society. Such efforts have generally relied on sales price as a proxy for social value.

Environmental releases are, of course, an incomplete indicator of actual environmental impact. The impact of any contaminant depends on its biochemical or physical properties as well as on a number of local or global conditions. For example, a given amount of an organic contaminant can have quite different effects on water quality depending on the receiving water's volume, status (quiescent or turbulent), composition (fresh or saline), and initial condition (e.g., pH, concentration of dissolved $O_2$). Therefore, while determining the mass of a release is critical to assessing environmental impact, this almost always provides a less than full accounting of the actual potential for environmental harm.

Operational metrics may also address product design, packaging, and transport. Given the conventional wisdom that 80 percent of a product's attributes are determined during the first 20 percent of the product development process, efforts to measure and improve environmental performance during the design phase can be particularly fruitful (DeLadurantey et al., 1996; Hoffman, 1997). Because

BOX 2-1
Examples of Operational Metrics for Evaluating
Environmental Performance

Materials
- Quantity of materials used per unit of product
- Quantity of water per unit of product
- Quantity of processed, recycled, or reused materials
- Quantity of packaging materials discarded or reused per unit of product
- Quantity of auxiliary materials recycled or reused
- Quantity of water reused
- Quantity of hazardous materials used in the production process

Energy
- Quantity of energy used per year or per unit of product
- Quantity of energy used per service or customer
- Quantity of each type of energy used
- Quantity of energy units saved due to energy conservation

Physical Facilities and Equipment
- Number of pieces of equipment with parts designed for easy disassembly, recycling, and reuse
- Number of freight deliveries, by mode
- Number of vehicles in fleet with pollution abatement technology
- Number of products that can be reused or recycled
- Average fuel consumption of vehicle fleet

Products
- Number of products introduced in the market with reduced hazardous properties

Wastes
- Quantity of waste per year or per unit of product
- Total waste for disposal
- Quantity of hazardous waste recycled
- Quantity of hazardous waste eliminated due to material substitution

Emissions
- Quantity of specific emissions per year
- Quantity of specific emissions per unit of product
- Quantity of waste energy
- Quantity of air emissions having ozone depletion potential
- Quantity of air emissions having global climate change potential

SOURCE: International Organization for Standardization, Annexes Testing Committee (1996).

many companies outsource a considerable portion of their activities, some attention is also focusing on the environmental performance of suppliers who provide raw, unprocessed, or unassembled production materials and even complete subassemblies.

Operational metrics have, thus far, received the majority of industry's attention. This is due in part to their close connection to regulatory compliance. The fact that these metrics often address aspects of production that are both tangible and within a company's ability to control also should not be overlooked.

## Management Metrics

Whereas operational metrics provide an indication of the present state of a company's environmental performance, management metrics furnish information on steps being taken to influence operations. Management metrics describe such things as the allocation of funds and labor, implementation of environmental programs and new environmental policies, environment-related legal expenses, environmental remediation activities, and the status of environmental information systems (Box 2-2). Metrics like these are designed to inform management

---

**BOX 2-2**
**Examples of Management Metrics for Evaluating**
**Environmental Performance**

- Number of achieved objectives and targets
- Number of pollution prevention initiatives implemented
- Number of levels of management with specific environmental responsibilities
- Number of employees whose job descriptions include environmental responsibilities
- Number of employees trained versus the number that need training
- Sales revenue attributable to a new product or a by-product designed to meet environmental performance objectives
- Degree of compliance with environmental regulations
- Number of resolved and unresolved corrective actions
- Number of or costs attributable to fines and penalties
- Progress on local remediation activities
- Operational and capital costs associated with a product's or process's environmental aspects
- Savings achieved through reductions in resource use, prevention of pollution, or waste recycling
- Environmental liabilities that may have a material impact on the financial status of the organization

SOURCE: International Organization for Standardization, Annexes Testing Committee (1996).

and support decision making on the expenditure of time, money, and manpower required to maintain or improve a company's environmental performance. Information on how and where corporate resources are allocated can help to identify problem areas and opportunities for improvement. Management indicators, particularly those related to budgeting (i.e., for capital, operations, and maintenance), may also help companies assess the worth of previous environmental investments. Projects and programs that show a positive return on investment provide incentive for continued efforts to improve environmental performance.

### Environmental Condition Metrics

In the final analysis, it is generally environmental condition metrics that are of greatest interest to industry and external stakeholders. Unfortunately, this is also the area in which the fewest number of robust metrics have been developed and implemented. Environmental condition metrics seek to provide information on the health of the environment and how it is changing. Ideally, these metrics would link specific industrial activities or emissions to environmental impacts. Establishing a causal relationship between pollutants and impacts, however, can be a complex undertaking. Nonetheless, there are a growing number of measures that are applied to environmental systems in an attempt to assess their relative health (Box 2-3).

Many of these measures have been derived from regulations or guidelines and are intended to maintain environmental standards that protect the health of humans and indigenous flora and fauna. Numerous measures of regulatory origin have been used to characterize water and air quality and soil and groundwater contamination in local or regional ecosystems. Environmental condition metrics have also been formulated for use at a global scale. Measures of atmospheric ozone and global mean temperature are examples of such indicators. These types of metrics are often proposed initially by the scientific community and later appropriated by governmental and nongovernmental organizations for use in assessing environmental health.

In recent years, increasing attention has been given to a broad range of metrics that seek to estimate the long-term sustainability of human activities. The definition of sustainability, particularly as might be applied within a corporate context, is still evolving, but some companies and researchers have begun to investigate how such principles might be practically implemented. This report explores some of these experimental efforts while also describing current practice in assessing environmental condition.

## CHARACTERISTICS OF A GOOD METRIC

The committee believes that an essential precondition for developing useful environmental metrics is a well-defined set of environmental goals or objectives.

---

**BOX 2-3**
**Examples of Environmental Condition Metrics for Evaluating Environmental Performance**

Air
  • Concentration of a specific contaminant in ambient air at selected monitoring locations
  • Odor measured at a specific distance from the organization's facility

Water
  • Concentration of a specific contaminant in groundwater or surface water
  • Turbidity measured upstream and downstream of a facility's wastewater discharge point
  • Dissolved oxygen in receiving water

Land
  • Concentration of a specific contaminant in surface soils at locations in the area surrounding the facility
  • Defined local area rehabilitated

Flora and Fauna
  • Quantity and quality of vegetation in a defined local area
  • Number of animal species in a defined local area
  • Population of a particular animal species within a defined distance of a facility
  • Quality of habitat for specific species in the local area

SOURCE: International Organization for Standardization, Annexes Testing Committee (1996).

---

Once an objective has been established, an effective metric will influence behavior by measuring an organization's progress toward achieving that objective. While environmental goals may vary across an industry and over time, *the vital attribute for any metric is that it shows a clear relationship to the desired objective.* In order for a metric to be useful, it also must be possible to obtain reliable measurements and data; the information gathered must be relatively unambiguous and readily understandable; and it must be communicated internally to those decision makers capable of affecting change or, alternatively, to interested external stakeholders.

In addition to these general properties, industry, public stakeholders (e.g., local communities, citizen groups), and commercial stakeholders (e.g., lenders, investors, insurers) attach importance to certain specific attributes.

### Characteristics Desired by Industry

Some characteristics of metrics, especially those tied to operational and management decisions, have particular value within company walls. In 1996 the

Tellus Institute and World Resources Institute (WRI) conducted an informal survey of environment, health, and safety professionals to identify the characteristics industry finds most useful in an environmental metric (White and Zinkl, 1996, 1997). Survey respondents represented 33 mostly large (annual revenues in excess of $1 billion) corporations from several different industrial sectors (e.g., chemicals, electric, gas, sanitary services).

Survey respondents attached particular importance to two characteristics: the ability to verify metrics and the ability to reliably compare measurements over time (Figure 2-1). The importance of verifiability is due in part to traditional concerns over compliance. The ability to defend the validity of data or measurements is paramount in efforts to demonstrate compliance with permits or with state and federal regulations. The importance attached to comparability over time indicates industry's interest in tracking improvements in environmental performance. Without such comparisons it is impossible to determine whether actual progress has been made. The importance of this characteristic may increase in the future if more companies begin to link incentives (i.e., salary, bonuses, etc.) to environmental as well as financial performance. This same line of reasoning may also explain the relatively high value survey respondents attached to the comparability of metrics within a company. Such comparability provides consistency across different business units in an organization. It also improves a company's ability to compare the environmental impacts of disparate processes or products.

Most of the characteristics cited by survey respondents were of primary relevance within companies themselves, but some industry respondents showed substantial interest in metric attributes more useful for external application. For example, the ability to account for the scale of corporate operations through some form of normalization (e.g., per unit produced, per dollar of sales revenue) was thought to be an essential or helpful metric property by 92 percent of those polled.

Eighty-two and 76 percent of respondents, respectively, cited comparability across companies within an industrial sector and comparability across industrial sectors as being either essential or helpful metric attributes. Finally, and perhaps a bit surprisingly, 26 of the 33 individuals polled said that public reportability would be either an essential or a helpful metric attribute. The number of these characteristics and the degree to which each is incorporated into the development of new metrics will presumably vary depending on process and industry.

Finally, it should be noted that companies also desire metrics that can be used in building relationships with external stakeholders. These stakeholders include those with a direct financial interest in the organization or facility (e.g., lenders, insurers, investors) as well as the communities in which companies operate. An open exchange of information on environmental performance can be an important step in building trust with these stakeholders, particularly local

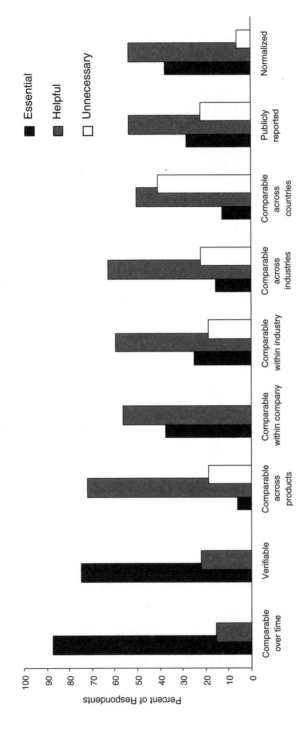

FIGURE 2-1 Relative importance of characteristics of environmental performance indicators. SOURCE: White and Zinkl (1996).

communities. Such public relations efforts can be important to avoiding negative media attention and more intensive regulatory scrutiny.

## Characteristics Desired by Public Stakeholders

As considered by the committee, external stakeholders are individuals and groups that generally have little direct financial interest in an industrial operation or sector. Individuals, public interest organizations, and community groups tend to be concerned about environmental performance information that relates to potential threats to human health. However, these stakeholders also show evidence of becoming increasingly aware of long-term issues related to ecosystem health and sustainability (Delphi Group, 1998). Human health, ecosystem health, and sustainability are all related at some level. Current capability to scientifically link each of these to various environmental loads is limited, however.

Although much work remains to be done in the area of human toxicology, some clear connections between human health and exposure to hazardous substances have been established. Methods of assessing ecosystem health are somewhat less developed, but substantial progress has been made over the past 25 years in establishing standards for maintaining a robust local or regional environment. By comparison, public interest in sustainability is relatively new, and, accordingly, there is still considerable debate over how the concept should be defined, measured, and addressed.

Most stakeholders seem to accept the notion that compliance with federal and state environmental regulations protects public health. Nonetheless, hazardous emissions and on-site storage of hazardous materials, even when fully compliant with existing regulations, are of considerable concern to local communities. These concerns become much more acute when regulatory violations occur. Local communities also tend to react quickly and forcefully to nuisance problems such as odors, noise, and dust, whether or not these emissions pose a health risk.

Public concerns over ecosystem and human health are often comingled when water quality, air quality, and soil or groundwater contamination are at issue. Tracking the amount of persistent or bioaccumulative substances in the environment, even if within regulatory limits, will continue to be an area of industrial activity that receives considerable public scrutiny. Although the U.S. Environmental Protection Agency's (EPA's) Toxic Release Inventory (TRI) provides some useful data, the aggregate nature of the measurements and the lack of information regarding relative impacts often leave the public wanting for details. Some companies have tried to overcome these limitations (Imperial Chemical Industries Group, 1996; Wright et al., 1998), but no impact weighting system has yet proved widely acceptable.

The need for systems for ranking or weighting toxic releases is a major issue in the metrics field, as evidenced by the number of researchers and nongovernmental environmental organizations that have undertaken their own efforts to disaggregate and rank pollutant emissions (usually based on TRI data). The EPA has also expended significant effort to devise a system for ranking pollutants (Hazard Ranking System) and provide guidelines on toxic thresholds (Ecotox Thresholds Software) (United States Environmental Protection Agency, 1999). The Environmental Defense Fund has recently established an online system that uses local maps to graphically link TRI emissions data to thousands of specific industrial sites throughout the United States. The website, called Scorecard, can disaggregate and weight emissions based on such characteristics as carcinogenicity. The website received over 1 million hits in its first 2 weeks, an indication of the public's considerable interest and demand for such information.

In addition to water, air, and soil quality, many public stakeholders express concern over the health of indigenous flora and fauna. In this context, measures of the number and type of various species being supported by an ecosystem have proved useful at both the local and the regional levels, while concern over endangered species often takes on global meaning. The public has also taken some interest in assessments of "nuisance" behavior caused by industrial activity occurring near or in an ecosystem (e.g., damaged aesthetic appeal, odors, noise, impaired visibility).

In recent years the public has begun to recognize the long-term environmental impacts of human activities and the need to assess these impacts. A number of investigations (including this one) are seeking to flesh out industry's role in measuring and adjusting its behavior in a manner that will improve sustainability. Measuring the use of renewable as opposed to depletable resources is one way some companies are trying to track more-sustainable activities. Likewise, some firms have begun to disaggregate and track energy sources (e.g., coal, oil, natural gas, solar) by carbon content or global warming potential to aid efforts to shift to less environmentally burdensome resources. Developing meaningful measures of sustainability is a major challenge for industry and society.

### Characteristics Desired by Financial Stakeholders

Stakeholders in the financial sector (e.g., banks, insurers, investment companies) are beginning to express some interest in corporate environmental performance. Many insurance and lending institutions have started to factor environmental information into their decisions regarding insurability, premiums, and loans. Much of this relatively new-found interest reflects the risks companies faced as a result of the 1980 Comprehensive Environmental Response, Compensation, and Liability Act (CERCLA, "Superfund"). CERCLA makes current and past contributors to environmentally hazardous sites financially liable for cleanup. This liability can be transferred to those who buy a contaminated site, significantly increasing the financial risk in any type of property exchange. As a result, almost all real estate transactions now require an environmental impact assessment. Such assessments are also commonly required before insurance companies will issue coverage.  CERCLA's "strict" liability provision is largely indifferent to whether the offending actions were legal or not at the time they occurred.  The law thus provides ample incentives for companies to reduce or eliminate activities (legal or otherwise) that may adversely affect the environment.  Unfortunately, this law has had the unintended consequence of encouraging some industrial and commercial developers to seek contaminant-free "greenfields" rather than use existing developed areas.

Beyond the insurance and lending sectors, there are indications that the investment community is now beginning to take some interest in the environmental performance of companies.  Investor attention to environmental issues is not widespread, although some research has posited a link between financial and environmental performance (Blumberg et al., 1996; Cohen et al., 1995; Hart and Ahuja, 1996).  Still, many within the investment community remain unconvinced.

The United Nations Development Programme recently produced a working paper (Gentry and Fernandez, 1996) containing results from a survey of selected Fortune 500 chief financial officers (CFOs) and securities analysts.  Respondents were asked to describe their use of environmental information in assessing the relative strength of companies (Box 2-4).  Lack of confidence in quantitative measures that link environmental performance to financial performance was clear (question 1).  However, in terms of qualitative evaluation, the environment appears to take on a somewhat larger role (question 2).  The value of qualitative factors seems to outweigh that of strictly numerical data in the view of both the CFOs and analysts (question 3).  Although it seems that environmental stewardship plays some role in assessing the economic health and investment potential of a company, the penalties for poor environmental performance appear much greater than the rewards for excellence (questions 4 and 5).  This may be due to the absence of any agreed-upon measures for determining superior environmental performance.  Regulatory compliance, on the other hand, is a firm lower bound by which corporate laggards can be judged.

Survey respondents' concern over compliance issues was clear when they were asked to rank environmental factors related to financial performance and management quality (questions 6 and 7).  Corporate efforts to use environmental performance for competitive advantage received mixed reviews (question 8).  While CFOs attach some value to ecoefficiency programs, analysts are either unconvinced or uninformed as to the cost-saving potential of such initiatives.  It seems that if ecoefficiency activities are ever to receive attention from the financial community, communication of the rationale and potential benefits of such efforts must be improved (question 9).  This underscores the importance of communicating environmental information to groups lacking extensive industrial experience.  If metrics are to be widely used by external stakeholders, particularly those in the financial community, they will need to be both useful and comprehensible.  The creation of transparent measures of performance may help overcome a number of actual and perceived barriers to industry incorporating envi-

BOX 2-4
Effects of Environmental Performance on Shareholder Value[a]

1. How Important are the following quantitative factors in your overall analysis of a company?

|  | Analysts | CFOs |
| --- | --- | --- |
| Sales | 3.3 | 3.1 |
| Return on equity | 3.7 | 4.6 |
| Margins | 4.4 | 4.1 |
| Earnings growth | 4.0 | 4.6 |
| Cash flow | 4.5 | 4.5 |
| Potential for industry growth | 3.7 | 3.6 |
| Potential to gain market share | 3.9 | 3.8 |
| Employee turnover | 2.4 | 2.9 |
| Research and development | 2.7 | 3.3 |
| Environmental spending | 1.9 | 2.8 |

2. How important are the following qualitative factors in your overall analysis of a company?

|  | Analysts | CFOs |
| --- | --- | --- |
| Quality of management | 4.7 | n.a. |
| Customer satisfaction | 3.7 | 4.6 |
| Employee satisfaction | 3.0 | 4.3 |
| Reputation in business community | 3.4 | 4.1 |
| Reputation among general public | 3.1 | 3.9 |
| Corporate environmental policy | 2.3 | 3.8 |

3. What is the importance of qualitative factors relative to quantitative factors?

| Analysts | CFOs |
| --- | --- |
| 3.4 | 3.8 |

4. Will investors pay a premium for a company with an exceptional environmental program?

|  | Analysts | CFOs |
| --- | --- | --- |
| Percent answering *yes* | 17 | 25 |

5. Will investors apply a discount to companies with poor environmental performance?

|  | Analysts | CFOs |
| --- | --- | --- |
| Percent answering *yes* | 70 | 75 |

*continued*

6. How important are the following environmental factors in evaluating a company's financial performance?

| | Analysts | CFOs |
|---|---|---|
| Compliance costs | 2.9 | 3.6 |
| Litigation costs | 3.0 | 3.5 |
| Cleanup liabilities | 3.1 | 3.8 |
| Waste disposal costs | 2.6 | 3.3 |
| Environmental improvement investments | 2.4 | 3.1 |
| Pollution prevention savings | 2.1 | 3.0 |
| Energy efficiency savings | 2.8 | 3.4 |

7. How important are the following environmental factors in evaluating a company's quality of management?

| | Analysts | CFOs |
|---|---|---|
| Spills, violations, accidents | 3.4 | 3.9 |
| Environmental policy | 2.9 | 3.5 |
| Environmental auditing program | 2.6 | 3.8 |
| Third-party environmental certification | 2.4 | 3.1 |

8. How important are the following environmental factors in evaluating a company's potential to gain competitive advantage?

| | Analysts | CFOs |
|---|---|---|
| Recyclability of products | 2.2 | 3.1 |
| Resource recovery/recycling program | 2.4 | 3.3 |
| Pollution prevention program | 2.9 | 3.3 |
| Waste reduction program | 2.3 | 3.1 |
| After-market product take-back | 1.8 | 3.0 |
| Ecolabeling of products | 1.9 | 3.0 |

9. How well do companies communicate the following?
*[Scale is from 1 (badly) to 5 (well).]*

| | Analysts | CFOs |
|---|---|---|
| Their rationale for short-term environmental spending (1–2 years) | 2.2 | 3.3 |
| Their rationale for long-term environmental spending (5–10 years) | 2.1 | 3.3 |
| The potential influence of environmental factors on future revenues and risks | 1.9 | 3.3 |
| The potential influence of environmentally driven commercial demands on competitive advantage | 1.9 | 3.1 |
| The difference between reactive environmental compliance and proactive environmental improvement investing | 1.7 | 2.8 |

*continued*

BOX 2-4 *Continued*

10. How significantly do the following hinder the incorporation of environmental factors in your evaluation?

|  | Analysts | CFOs |
|---|---|---|
| Not a financial issue | 2.9 | 2.3 |
| Financial impacts are relatively insignificant | 3.4 | 2.5 |
| Financial impacts are outside the time frame applied by analysts | 3.0 | 3.3 |
| Available environmental data are unreliable | 3.1 | 3.7 |
| Lack of quantifiable data | 3.5 | 3.8 |
| Lack of tools for quantifying environmental data | 3.4 | 3.7 |
| Inadequate sources of useful data | 3.7 | 3.5 |
| Gaps between financial and environmental language | 2.8 | 2.7 |
| Lack of access to company environmental personnel | 2.2 | 2.7 |

[a] Except for questions 4, 5, and 9, data represent mean responses, where importance was ranked on a scale of 1 (least important) to 5 (most important). CFOs were asked 16 questions and analysts 20. The wording of some questions was slightly different in the two surveys.

SOURCE: Gentry and Fernandez (1996).

ronmental factors into decision making (question 10). Although a small segment of the investment community does make use of environmental performance data, the large majority of investors gives such information little attention.

## USE OF ENVIRONMENTAL METRICS IN INDUSTRY DECISION MAKING

While it is apparent that a vast array of environmental parameters is being measured, tracked, and reported, little is known about the role this information plays in the day-to-day activities of companies. The value of a metric (or suite of metrics) relates directly to the number and nature of corporate decisions it influences.

Respondents of the joint Tellus Institute–WRI survey (White and Zinkl, 1997) were asked to describe and rank the types of decisions that were most affected by environmental performance metrics (Table 2-1). As might be expected, regulatory compliance topped the list: Over 60 percent cited this as the number one area of business decision making affected by metrics. Another 24 percent ranked compliance decisions second or third in importance. Strategic business planning ranked next in terms of importance, with 15 percent of those

TABLE 2-1   Business Decisions Most Affected by Environmental
Performance Metrics

| Business Decision | Percent of Respondents Ranking it First | Percent of Respondents Ranking it First, Second, or Third |
|---|---|---|
| Purchasing | 9 | 18 |
| Product design | 3 | 15 |
| Strategic planning | 15 | 64 |
| Regulatory compliance | 61 | 85 |
| Employee compensation | 0 | 0 |
| Marketing | 0 | 3 |
| Research and development | 0 | 12 |
| Investment decision making | 0 | 30 |
| Internal benchmarking | 3 | 30 |
| External benchmarking | 3 | 21 |

SOURCE:  White and Zinkl (1996).

polled placing it first and another 49 percent ranking it either second or third. Environmental metrics would therefore seem to have some impact on corporate investment decisions, but their utility for decision making in an array of other core business areas appears to be either limited or underutilized.

Given this, it is interesting to note how respondents ranked the overall importance of environmental criteria in corporate decision making (Figure 2-2). Sixty-seven percent "almost always" made use of environmental information when weighing decisions. When these responses are considered together with those given to the question of whether "adequate metrics" are currently available (Figure 2-3), it seems that a considerable gap exists between the type of environ-

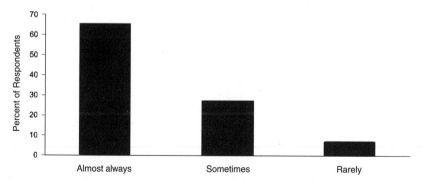

FIGURE 2-2   Frequency of use of environmental criteria in corporate decision making. SOURCE:  White and Zinkl (1996).

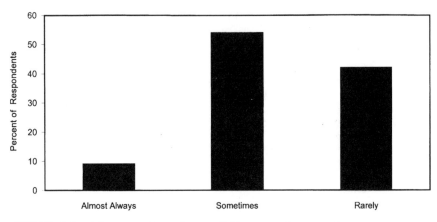

FIGURE 2-3 Adequacy of metrics available to evaluate products or processes. SOURCE: White and Zinkl (1997).

mental information required and that being supplied. It may simply be that metrics are used primarily for compliance because that is the area in which most metrics exist. There seems to be a considerable unmet demand for environmental information; substantial rewards might accrue to those able to collect and make use of it.

## REFERENCES

Blumberg, J., A. Korsvold, and G. Blum. 1996. Environmental Performance and Shareholder Value. Conches-Geneve, Switzerland: World Business Council for Sustainable Development.

Cohen, M., S. Fenn, and J. Naimon. 1995. Environmental and Financial Performance: Are They Related? Vanderbilt University, Owen School of Business. Nashville, Tenn.: Investor Responsibility Research Center.

DeLadurantey, C.E., R.J. Kainz, and M.H. Prokopyshen. 1996. Environment, health, and safety: A decision model for product development. Paper presented at the SAE International Congress and Exposition, Detroit, February 26–29.

Delphi Group. 1998. A Business Guide: Environmental Performance and Competitive Advantage. Ontario: Queen's Printer for Ontario.

Ditz, D., and J. Ranganathan. 1997. Measuring Up: Toward a Common Framework for Tracking Corporate Environmental Performance. Washington, D.C.: World Resources Institute.

Gentry, B.S., and L.O. Fernandez. 1996. Valuing the Environment: How Fortune 500 CFOs and Analysts Measure Corporate Performance. United Nations Development Programme (UNDP), Office of Development Studies, Working Paper Series. New York: UNDP.

Hart, S.A., and G. Ahuja. 1996. Does it pay to be green? An empirical examination of the relationship between emission reduction and firm performance. Business Strategy and the Environment 5:30–38.

Hoffman, W.F. 1997. A tiered approach to design for environment. Paper presented at the Conference on Clean Electronic Products and Technology, Institute for Electrical and Electronics Engineers, Edinburgh, Scotland.

Imperial Chemical Industries Group (ICIG). 1996. Annual Report on Safety, Health, and Environmental Performance. London: ICIG.

International Organization for Standardization, Annexes Testing Committee. 1996. ISO 14031 Environmental Performance Evaluation. Final ATC Report Edition. U.S. SubTag 4. New York: American National Standards Institute.

United States Environmental Protection Agency. 1999. Ecotox Threshold. Available online at http://www.epa.gov/oerrpage/superfnd/web/oerr/r19/ecotox/index.html. [January 7, 1999]

White, A., and D. Zinkl. 1996. Corporate Environmental Performance Indicators: A Benchmark Survey of Business Decision Makers. Boston: Tellus Institute.

White, A., and D. Zinkl. 1997. Green Metrics: A Status Report on Standardized Corporate Environmental Reporting. Boston: Tellus Institute.

Wright, M., D. Allen, R. Clift, and H. Sas. 1998. Measuring corporate environmental performance: The ICI environmental burden system. Paper presented at the NAE Workshop on Industrial Environmental Performance Metrics, January 28–29, Washington, D.C.

# PART II

# The Industry Studies

# 3

# Guide to the Industry Studies

Compared with small or medium-sized enterprises, large organizations often have greater resources to devote to improving environmental performance. With this in mind, it should be noted that this report carries the bias of the study committee, the bulk of whose members are presently, or have been, employed by larger firms.[1] The lessons and recommendations presented here may not apply to all industries or even all firms within the sectors analyzed. However, the committee believes that despite these limitations the results of this study provide a framework that will be useful to a wide range of industries as they seek to improve their development and use of industrial environmental performance metrics.

The four sectors studied—automotive, chemical, electronics, and pulp and paper—represent very different parts of the industrial system, as illustrated by the simple materials flow model in Figure 3-1. The automotive industry purchases both manufactured parts and refined materials, which it uses to manufacture vehicles. Some waste or nonproduct output that results from the manufacturing process is recycled directly into the input stream. Most vehicle parts and materials are reused or recycled at the end of their useful lives, some by the auto industry itself, others by industries separate from the original equipment (new-vehicle) manufacturers. The chemical industry primarily refines input materials and manufactures products and wastes. The industry also uses some of its product and

---

[1]With one exception, at the time of the study industry members of the committee worked for manufacturing firms with 1996 annual revenues in excess of $5 billion.

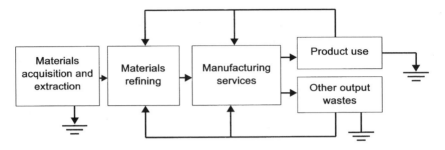

FIGURE 3-1    Analytic framework for the industry case studies.
NOTE:  The electrical ground symbol indicates where materials, wastes, or products are disposed of or lost to the environment.

nonproduct output materials as input to other processes.  The electronics industry buys semifinished parts, performs some further refinement, manufactures products, and produces nonproduct output and waste, some of which can be recycled. Semiconductors, for example, are incorporated into a wide range of finished electronics goods such as computers, whose disposal is becoming of increasing concern as more and more enter the waste stream.  The pulp and paper industry often grows, harvests, and refines its own materials.  It also uses some nonproduct output to generate energy and recycles significant fractions of both pre- and postconsumer products.

The analytic framework described in Figure 3-1 focuses primarily on materials flows and assumes that energy is integral to all aspects of operations.  Environmental metrics examined in this context are applied to the production process and sometimes downstream to the end of a product's life, but they generally are not applied upstream to raw materials acquisition (except in the case of the pulp and paper industry, where there is an obvious link between raw material and product).

As knowledge about industrial impacts on the environment has improved, companies have been asked to begin to look beyond these limits.  The boundaries of concern have begun to expand to include more of the product life cycle and the environmental performance of suppliers.  Also emerging are societal concerns related to such issues as biodiversity and sustainable development, which traditionally have not shown up on industry's radar.  Relating these concerns to typical industrial operations will be difficult.  It will require the use of weighting schemes, long-term forecasting, and subjective judgments, all of which are presently subject to considerable uncertainty.  Current practice still displays elements of the type of thinking prevalent 30 years ago, when environmental considerations were essentially independent of business decisions.  This is beginning to change, however, both as a result of the competitive advantages being realized and the long-term issues (e.g., ecosystem impacts, sustainability, and associated socioeconomic issues) that increasingly dominate discussions on environmental policy.

The following industry studies examine environmental performance metrics in three categories: resource use, environmental burden, and human health and safety. Corporate human health and safety metrics, while sometimes related to environmental metrics, are not the focus of this report. Nonetheless, because health and safety metrics are often tracked and reported by the same corporate unit responsible for environmental compliance, they are included in the summaries of metrics currently in use.

The diverse industries examined in this report serve to illustrate different aspects of metrics and their potential uses. The automotive case study explores the full spectrum of operational and product-related metrics used by an industry producing a finished consumer product. The chemical sector produces refined raw materials that are used primarily as feedstocks in other industries. This characteristic has pushed the industry to explore the use of weighting and normalization in its metrics. The analysis of the electronics sector details the use of metrics in the manufacture of semiconductors, which are incorporated into a wide range of finished goods. Addressed in this case study is the question of what measures of environmental performance are relevant to the final assembled product. The pulp and paper sector maintains some control over many stages of product life cycle, from raw materials acquisition to end-of-life processing (i.e., recycling). This presents the industry with the challenge of being more comprehensive in its metrics.

All four case studies present metrics in terms of an input-output flowchart and a matrix relating the metrics to manufacturing operations and products. Each provides basic background, including a description of the industry's production processes. Metrics currently in use are discussed in detail, as are the challenges to and opportunities for improving current metrics or developing new ones. The automotive study includes a number of metrics that are common to the other three sectors. Therefore, a full description of many of these common measures will not be undertaken for each industry.

# 4

# The Automotive Industry

## BACKGROUND

The U.S. automotive industry is composed of three major U.S.-based manufacturers (Chrysler, Ford, and General Motors),[1] several non-U.S.-based (transplant) vehicle assemblers, and a vast network of parts and components suppliers. Collectively, the industry produces and sells approximately 15 million cars and light trucks each year. Total sales in 1997 were nearly $500 billion, and total employment was nearly 1 million. Manufacturing facilities include small specialty-parts plants, large foundries and engine and transmission plants, and vehicle assembly plants, which employ thousands of people and produce several hundred thousand vehicles per year.

## Automobile Manufacturing

The industry's main products are automobiles, light and heavy trucks, and sport utility vehicles. These are produced using various casting, stamping, molding, welding, painting, and assembly processes. Each operation poses a unique set of environmental challenges. In addition, while automobile manufacturers do not directly recycle vehicles, their products, at the end of life, are extensively recycled through independent dismantlers and shredders (Figure 4-1). A portion

---

[1]The merger of the German conglomerate Daimler Benz and Chrysler took place after the bulk of this study had been completed. Although this reduces the number of U.S.-based automakers from three to two, the effects of the merger, particularly over the next several years, should have little bearing on the observations and recommendations contained in this report.

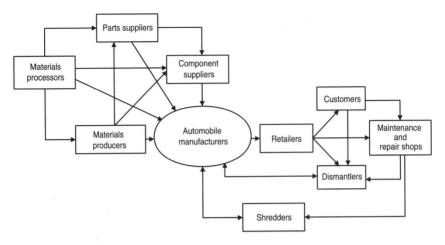

FIGURE 4-1  Economic players in the automotive sector.
SOURCE:  Adapted from Fischer and Schot (1993).

of the recycled parts and materials becomes inputs to various automotive processes; the rest is used elsewhere in the economy. Each year approximately 10
million automobiles, buses, trucks, and motorcycles are processed by dismantlers,
who supply 37 percent of the nation's ferrous scrap (American Automobile Manufacturers Association, 1994).

### Drivers of Environmental Performance Improvements

The life cycle of a typical automobile and the various processes associated
with different parts of the cycle are shown in Figure 4-2.  Within this life cycle,
efforts to improve environmental performance are focused on manufacturing
processes, product use, and end-of-life recycling.  In manufacturing, attention is
paid to the solid-, liquid-, and gas-phase emissions from operations, as well as to
materials and energy usage;  product-related concerns focus primarily on exhaust
emissions and energy use.

Regulation is the primary driver of environmental change in the industry.
Federal regulations that affect automobile manufacturing include the Clean Air
Act (CAA), the Clean Water Act (CWA), the Resource Conservation and Recovery Act (RCRA), Superfund Amendments and Reauthorization Act (SARA), and
the Pollution Prevention Act of 1990.  The primary metrics used are derived from
these regulations, their amendments, and other negotiations resulting from proposed rule-making.

Environmental progress reported by the automobile companies relates primarily to these regulations, government-initiated voluntary efforts such as the

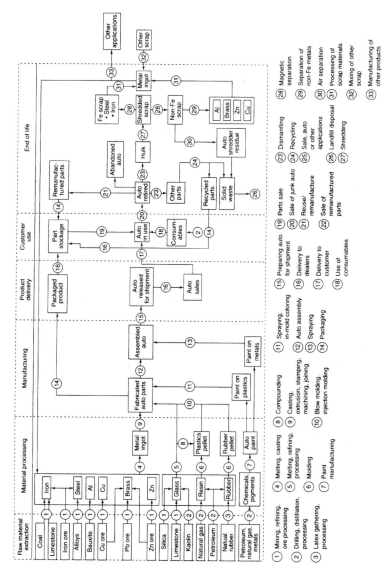

FIGURE 4-2 The life cycle of the automobile and the processes that occur during that cycle. The processes listed at the bottom of the chart are keyed by number to the life-cycle steps shown in the flow diagram.

SOURCE: Graedel and Allenby (1997). Copyright ©, 1996, Lucent Technologies. Used by permission.

U.S. Environmental Protection Agency's (EPA) 33/50 program,[2] and regional initiatives such as the Great Lakes persistent toxics (GLPT) program.[3] Automakers also undertake pollution prevention efforts to improve water-, materials-, and energy-use efficiencies. Details of specific actions taken and the cost savings that have accrued are well documented by the Michigan Department of Environmental Quality (1998a).

Competitive pressures, particularly from overseas manufacturers, along with the advent of information technologies and new management techniques have also prompted dramatic changes in automotive design and manufacturing processes. Total quality management, just-in-time inventory control, concurrent engineering, and lean-production techniques are some of the approaches that have been implemented by domestic manufacturers and suppliers to maintain competitiveness. Many of these initiatives have minimized inputs during production and led to cleaner production as well. Life-cycle management (Kainz et al., 1996) and design for environment practices (Ford Motor Company, 1998) are beginning to be used in the industry to meet regulatory and internal environmental goals.

The automotive industry's product—the vehicle—is also heavily regulated. Following the Organization of Petroleum Exporting Countries oil embargo that led to a tripling of oil prices in the early 1970s, the Energy Policy and Conservation Act (EPCA) of 1975 was passed. The law introduced minimum corporate average fuel economy (CAFE) standards for cars and light trucks. CAFE standards are calculated for car and light truck categories for each producer's fleet. Producers are penalized for each mile-per-gallon deficiency per vehicle, although credits for surpassing the standard can be earned. Until enactment of EPCA and CAA, increased horsepower and performance were obtained by using larger engines (National Research Council, 1992). This legislation, and several other laws that followed, made the production of lighter-weight vehicles and smaller engines with lower exhaust emissions an industry goal. Through the 1980s and early 1990s, increased computerization of automobile functions, introduction of

---

[2]EPA's 33/50 program (also known as the Industrial Toxics Project) is a voluntary pollution reduction initiative that targets releases and off-site transfers of 17 high-priority toxic chemicals. Its name is derived from its overall national goals—an interim goal of 33 percent reduction by 1992 and an ultimate goal of a 50 percent reduction by 1995, with 1988 as the baseline year. The 17 chemicals are from EPA's Toxic Release Inventory. They were selected because they are produced in large quantities and subsequently released to the environment in large quantities; they are generally considered to be very toxic or hazardous; and the technology exists to reduce releases of these chemicals through pollution prevention or other means. Although the goals have been met (a 40 percent reduction was achieved by 1992, and 50 percent reduction was reached ahead of schedule in 1994; United States Environmental Protection Agency, 1999), companies continue to track these 17 chemicals.

[3]The GLPT program is a partnership between industry and government to reduce the use of 65 toxic chemicals identified as significantly impacting the Great Lakes.

advanced materials, and a redesigned internal combustion automobile engine led to a significant decrease in engine size (over 1,500 cubic centimeters from 1974 to 1992) with no loss in horsepower (except for a dip in power in the late 1970s and early 1980s; National Research Council, 1992).

On another front, the industry is (and continues to be) challenged by regulatory demands for alternative fuels. For example, in response to the "energy crisis" of the 1970s and 1980s, several fuel alternatives to petroleum were developed. More recently, the Energy Policy Act of 1992 included several reformulated-fuel mandates aimed at lowering automobile hydrocarbon and carbon dioxide emissions.

Regulatory and competitive pressures have also resulted in several alternative-vehicle initiatives, such as the Partnership for a New Generation of Vehicles (PNGV). PNGV is a collaborative research and development program between the U.S. government and the U.S. Council for Automotive Research (USCAR), whose members are Chrysler, Ford, and General Motors. Its aim is to develop vehicles with fuel efficiency of up to 80 miles per gallon that will cost no more to own and operate than current comparable vehicles (e.g., the 1994 Chrysler Concorde, Ford Taurus, and Chevrolet Lumina; United States Department of Commerce, 1995). It is unclear if life-cycle assessments of the environmental impacts of the 80-miles-per-gallon cars will show the vehicles to be environmentally superior. For example, the new materials required will make recycling more difficult and less economical.

## CURRENT USE OF ENVIRONMENTAL PERFORMANCE METRICS

Environmental performance metrics have emerged in the automobile industry in response to regulation and to take advantage of opportunities to improve efficiency. The metrics are summarized in Figure 4-3. Manufacturing-related metrics allow companies to track material inputs to gain the maximum material-use efficiencies, pay attention to energy and water used in manufacturing, and track emissions from manufacturing operations. Product-related metrics relate to fuel economy and tailpipe emissions of hydrocarbons (HCs), oxides of nitrogen ($NO_x$), and carbon monoxide (CO). In addition, the industry tracks vehicle recycling.

### Manufacturing Metrics

Environmental metrics in auto manufacturing focus on waste and emissions and efforts to control them. Several sources of information are available, including Toxic Release Inventory (TRI) reporting (as required under the Emergency Planning and Community Right-to-Know Act of 1986, a part of SARA Title III), corporate annual reports, and reports from various voluntary partnerships. These, however, do not provide a complete picture of the environmental performance of

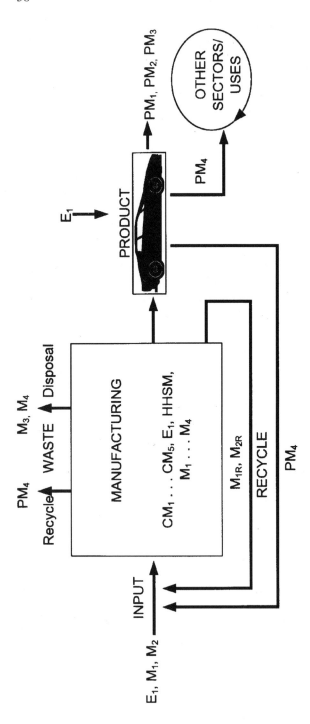

FIGURE 4-3 Environmental performance metrics in automobile production.

NOTE: $CM_1$ = Manufacturing emissions (other than TRI, 33/50, GLPT, and SARA Title II chemicals); $CM_2$ = Toxic Release Inventory chemicals; $CM_3$ = 33/50 chemicals; $CM_4$ = Great Lakes persistent toxics; $CM_5$ = SARA Title III chemicals; $E_1$ = Energy; $M_1 + M_{1R}$ = Parts, components, raw materials, and recycled materials; $M_2 + M_{2R}$ = Water and recycled water; $M_3$ = Packaging waste; $M_4$ = Solid waste (excluding packaging); $PM_1$ = Tailpipe HC, $NO_x$, and CO emissions; $PM_2$ = Evaporative emissions; $PM_3$ = Tailpipe $CO_2$ emissions; $PM_4$ = Recycled material from manufacturing process and product; HHSM = Human health and safety.

the industry.  Waste and emissions from this sector are distributed among suppliers as well as the Big Three manufacturers.  The large supplier base makes aggregating environmental metrics very difficult.  In general, however, the sector relies most on metrics to track wastes and emissions and metrics to manage materials, energy, and waste flows.

## Waste and Emissions

The various gross inputs to and outputs from automotive manufacturing processes are shown in Figure 4-4.  Nonproduct outputs include waste material, some of which is reused or recycled, and liquid or gaseous emissions.  Wastes and emissions are measured and reported in a variety of units (e.g., total quantities generated per year, quantities per unit product).  Reporting of TRI emissions provides the most common metric in the industry.  The TRI contains specific information about the release and transfer of toxic chemicals.  Transferred chemicals are those that are geographically or physically separate from a facility but still under its control.  More than 576 chemicals and 28 chemical categories were included in the 1997 TRI.  Any industrial facility with at least 10 full-time employees and that manufactures or processes 25,000 lbs. or uses 10,000 lbs. of a listed chemical has to report its emissions.  Since 1988, Chrysler, Ford, and GM together have reduced TRI releases by 53 percent, when normalized against production volume (American Automobile Manufacturers Association, 1998).  Box 4-1 provides an example of how process changes resulted from efforts to reduce toxic emissions.

Other voluntary efforts, such as the EPA's 33/50 and GLPT programs have similarly led to tracking of specific chemicals with the goal of reducing their use or emission.  In both of these efforts, similar reductions have been achieved.  Since 1988 the Big Three automakers have reduced the emission of 33/50 chemicals by over 60 percent on a per-vehicle normalized basis (American Automobile Manufacturers Association, 1998).  Trends in the release of GLPT chemicals (Figure 4-5) suggest that since 1991 there has been a reduction in aggregate releases (when zinc is excluded).  The anomaly for zinc is due to foundries recycling zinc-galvanized metal, which accounted for over 50 percent of all GLPT substance released.  Zinc releases are the result of increased recycling of galvanized steel, which has been used for body-panel corrosion protection.  When normalized by vehicle production volumes, overall releases of GLPTs have decreased by 9 percent since the GLPT program began in 1991.  When zinc releases from the foundries are excluded, the industry boasts a 54 percent decrease in GLPT emissions since 1991.  The goals of reducing TRI, 33/50, and GLPT chemicals have been mainly achieved through specific pollution prevention actions, process improvements, and recycling.  These efforts have been documented in numerous automotive industry case studies reported to the Michigan Department of Environmental Quality (1998c).

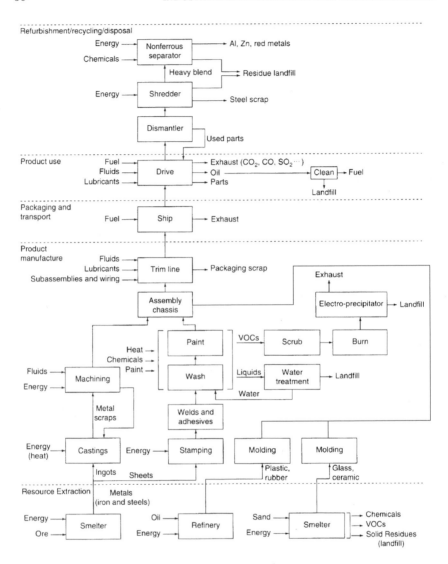

FIGURE 4-4  The life cycle of a typical automobile.  The life-cycle flow is from bottom to top.  Materials and energy inputs enter from the left; residues leave to the right.
NOTE: VOCs = volatile organic compounds.
SOURCE:  Graedel and Allenby (1997). Copyright ©, 1996, Lucent Technologies. Used by permission.

BOX 4-1
Process Change Eliminates the Use of TCE

Ford's Climate Control Division makes aluminum heat exchangers, such as radiators, heater cores, condensers, and evaporators. In the traditional process, trichlorethylene (TCE) is heated and used to degrease the very thin aluminum parts that are used to make the heat exchanger. After cleaning, the parts are assembled and brazed together as a coherent and leak-free unit. Although the degreasing process includes a TCE vapor collection system, some TCE remains on the high-surface-area parts and evaporates outside of the process equipment. A significant percentage of all the chlorinated solvents released annually by the company is due to this evaporation.

One alternative that appeared to have potential for replacing the TCE in this process was the use of a detergent and aqueous solution (water wash) that would not etch or damage the aluminum parts. A variety of detergents were tested. The two best-performing classes of detergents were then used in low-volume trials. At the same time, a design for a detergent and aqueous system was developed. With assistance from a supplier, an enclosed-water-spray system was chosen, in which the parts were moved through the spray areas by a belt feeder. The washer had three sections: a prewash for easy-to-remove oil, detergent wash to loosen and remove oil attached to the part surface, and a water rinse.

A low-volume aqueous pilot evaluation proved that the detergent alternative was compatible with current and future braze processes, and the system is now being used to reduce the company's dependency on TCE.

SOURCE: Michigan Department of Environmental Quality (1998b).

The largest point-source emissions in the automotive industry are volatile organic compounds (VOCs) used as paint solvents. Fifty solvents found in paints and adhesive solvents are among the 189 hazardous air pollutants regulated under Title III of the Clean Air Act Amendments of 1990. VOC emissions from these solvents occur during application, curing, and equipment cleaning operations. Several innovative paint technologies aimed at reducing the VOC burden associated with conventional solvent-base paint are emerging. Table 4-1 shows the goals, metrics used, and results of research conducted by the USCAR Low Emission Paint Consortium and several other cooperative research and development programs.

## Resource Use

Auto industry efforts to more efficiently use materials and energy have been driven by opportunities to reduce costs. Quality and just-in-time practices have targeted all materials used in the manufacturing process for efficiency improve-

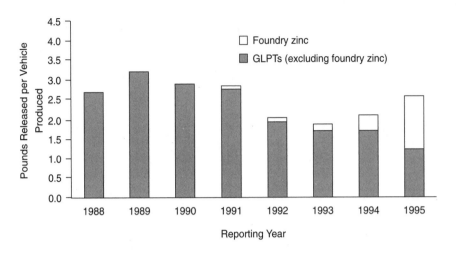

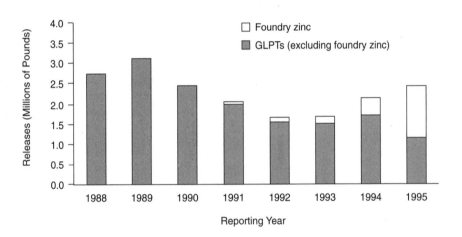

FIGURE 4-5 Great Lakes persistent toxic substance reportable production normalized and total releases for facilities belonging to the American Automobile Manufacturers Association. SOURCE: American Automobile Manufacturers Association (1998).

ments, including toxic materials covered under TRI such as asbestos, mercury, and lead. Metrics that may be used to track gains in resource-use efficiency include dollars saved, pounds of materials used per year, or kilowatt-hours consumed per vehicle produced.

In addition to improving resource efficiency, the industry is preventing pollution through the substitution of new parts or processes, a practice that requires sophisticated analysis and decision making. For example, Chrysler uses an approach it calls life-cycle management (LCM). LCM is used to evaluate costs for process changes (or alternative parts or components) together with environmental, occupational health and safety, and recyclability factors that are not considered in a traditional business analysis. Using LCM, Chrysler developed an underhood lighting system switch that eliminated mercury. Although the purchase price of the new switch was greater than that of the mercury-containing switch, its life-cycle costs were less (Box 4-2).

Energy or fuel appears as an input at every stage in the manufacture of vehicles (Figure 4-4). Energy is used to refine and process the raw materials, make the parts and components, assemble the vehicle, and deliver the product to showrooms. A recent life-cycle inventory estimated that about 164,000 MJ are used to produce a generic car (United States Automotive Materials Partnership, Life-Cycle Assessment Special Topics Group, 1997). This is the energy used in the production of materials and the manufacture of the vehicle only, not in vehicle use.

Total water use by the industry is also tracked, as is water use normalized according to liters per vehicle produced. Figure 4-6 reveals that water usage is heaviest during manufacturing and in processes that require a significant amount of cleaning. Concern over water use is often closely linked to local availability of this resource; the industry is making efforts to reduce water use across the board (Box 4-3).

## Reuse, Recycle, and Disposal

Solid waste associated with a typical automobile (1,370-kg car) is shown in Figure 4-7. These and other wastes have been the target of industry reuse and recycling efforts. Typical recycled quantities for one company are shown in Table 4-2. Reuse and recycling are important in ongoing efforts to optimize, as cost effectively as possible, energy use and material life cycles.

In general, reuse and recycling occur together, and the metric used is the amount of solid waste generated. By tracking this value over time, trends can be monitored (Figure 4-8). This metric can be expressed in tons per vehicle for both hazardous and nonhazardous wastes. Reductions can be expressed in dollars saved or in percent reduction achieved. Lists of material types that are recycled, expressed in number of pounds per vehicle, are maintained by automobile makers. This provides companies with the opportunity to track and report on the recycled content of their products.

Reuse by suppliers has also been encouraged throughout the industry. Working with suppliers, auto companies have reduced the costs of their own solid waste management and the costs of supplier packaging by requiring that materials delivered to plants be packaged in returnable dunnage. Progress in reducing packaging waste is shown in Figure 4-9. Other areas that show potential for reuse are also being explored, as illustrated by the initiative to reuse plastic (Box 4-4).

TABLE 4-1  Automotive Cooperative R&D Programs and Associated
Environmental Performance Metrics

| Program | Partners | Objective |
|---|---|---|
| USCAR Low Emission Paint Consortium | Chrysler, Ford, and General Motors in conjunction with paint and equipment suppliers. | To conduct joint R&D programs to reduce or eliminate solvent emissions from automotive painting systems and to accelerate the availability of low-emissions painting technology. The initial focus is on powder-painting technology. |
| USCAR Vehicle Recycling Partnership | Chrysler, Ford, and General Motors with Argonne National Laboratory through a cooperative research and development agreement. | To conduct cooperative R&D programs to increase the recyclability of automobiles; to promote design for recyclability in concert with dismantlers, reprocessors, and shredders. |
| USCAR Automotive Material Partnership Life-Cycle Inventory Program | Chrysler, Ford, and General Motors in cooperation with the Aluminum Association, American Iron and Steel Institute, and American Plastics Council. | To generate an environmental inventory for the entire life cycle of the generic family sedan (i.e., material production, parts manufacture, vehicle assembly, operation, maintenance, and disposal/recovery) by collecting data from the materials industries and automobile manufacturers plants. |
| Casting Emissions Reduction Program | U.S. Department of Defense, Chrysler, Ford, and General Motors with participation from the EPA, California Air Resources Board, and American Foundrymen's Society. | To improve and/or develop clean materials and processes in foundry technologies. |

Improvements such as these are captured in metrics that measure specific or total
material use.

## Product Metrics

Environmental performance metrics for the automotive industry's products,
as for its manufacturing process, have been driven largely by regulations and
focus on emissions, energy use, and recyclability.

| Environmental Performance Metrics | Status |
| --- | --- |
| Reductions in primary painting emissions, including volatile organic compounds (VOCs), solid waste, and water use. | A powder paint production facility has been constructed, and vehicles from Chrysler, Ford, and General Motors are being painted with powder clear coat at typical assembly line speeds. |
| Recycling of vehicles and vehicle parts. | Design for recyclability guidelines has been adopted by all the vehicle manufacturers based on dismantling studies completed at the center; pilot programs are under way to recycle polyurethane from seats, thermoplastic olefins from bumpers, and nylon from carpeting. |
| Air emissions; water and solid wastes across all life-cycle stages; energy consumption on a common basis between manufacturing and use of the product. | Quantitative inventory data are assembled for a large number of environmental performance metrics related to air, water, solid waste, resource use, and energy consumption; inventory results are reported for each of the life-cycle stages, and simulations are carried out to test the sensitivity of the results to the major assumptions and key parameters. |
| Reducing gas-phase emissions by 50% through the use of new materials; stretch goal of near-zero effect on the environment as measured by gaseous and particulate matter emissions. | Baseline environmental emissions have been established from current casting operations; pilot foundry has been constructed and validated; testing of new casting materials is under way in pilot foundry. |

## Emissions

Vehicle tailpipe emissions of HCs, CO, $NO_x$, and particulate matter have a long history of regulatory-driven environmental metrics. More recently, concerns have arisen about carbon dioxide ($CO_2$) emissions, particularly related to global climate change. $CO_2$ emissions are linked to energy use. Thus far, controls on energy use in automobiles have been set by CAFE standards. In the United States the vehicle emissions metric for cars and light trucks is mass

---

**BOX 4-2**
**Life-Cycle Management Study on the Impact of**
**Mercury-Free Switches**

To evaluate the impact of replacing a mercury switch with a mercury-free alternative for its vehicles' underhood lighting systems, Chrysler undertook a life-cycle management study. Mercury was targeted because its environmental impacts were gaining more and more attention. A dozen or so alternative designs for underhood convenience lighting systems were examined for feasibility and performance.

The study found that the total life-cycle management cost of the mercury-containing convenience lighting package is greater than that of the mercury-free alternative packages. The purchase price difference was $0.11 in favor of the current mercury switch; however, the relative cost savings when a total life-cycle analysis was conducted indicated a $0.12 advantage for the mercury-free alternative.

SOURCE: Michigan Department of Environmental Quality (1998d).

---

(grams) released per mile, as determined by test conditions specified by EPA regulations. Emissions from heavy-duty vehicles are also determined by EPA test procedures, but the metric is mass released per unit of engine power generated.

Emissions control technology is an important component of powertrain systems, and considerable research and development underlies the current systems. Regulatory requirements for reducing HC emissions are shown in Figure 4-10, along with proposed future reductions, such as those required to meet the strin-

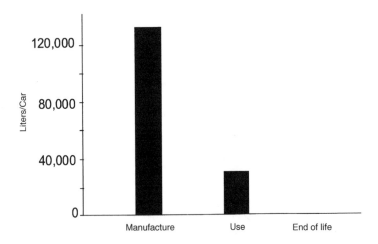

FIGURE 4-6  Water use over the life of a car.  SOURCE: Lee et al. (1997).

BOX 4-3
Reducing Water Use Through Innovation

Phosphating is the process of submerging completed metal auto bodies in a heated acidic solution that is rich in phosphate. The solution cleans and microscopically etches the metal surface so that paint will adhere to it. As soon as the auto body emerges from the phosphate, it is sprayed with water to remove the phosphating solution. Phosphate removal is critical to controlling the degree of etching and to avoid contaminating the anticorrosion chemicals into which the body is next submerged.

Because it is critical to remove all of the residue from each auto body, the water spray system used at the General Motors Midsize/Luxury Car Division's Fairfax Assembly Plant, located in Kansas City, Kansas, was designed to run whenever the phosphate system was running. The original design allowed the spray water to continue to flow during the lunch period, during the plant's two 23-minute breaks each day, and during unscheduled breaks in production. Further investigation revealed that the spray also operated during the 4-hour period prior to production each day that was required to bring the phosphate system up to operating temperature and for about an hour after the last auto body of the day moved through the spray.

To stop this wastage, GM designed and installed a photoelectric cell connected to a timer to start and stop the flow of rinse water. Now, if the photoelectric cell does not "see" an auto body pass by within a preset time interval, the valve is turned off and the water flow is halted. The single valve conserves approximately 6.5 million gallons of water annually and saves the facility approximately $33,000 (the costs of purchasing the water, treating it in the plant's wastewater treatment facility, and discharging it).

SOURCE: Michigan Department of Environmental Quality (1998d).

gent California requirements for ultra-low-emission vehicles and super-ultra-low-emission vehicles.

More recently, pollution prevention efforts have targeted evaporative emissions (i.e., fuel vapors that are produced in the fuel tank and in the fuel delivery system). These are measured in terms of evaporative emission rates.

## Resource Use

Energy consumed during a vehicle's service life is a measure of the product's performance. The energy consumed over the lifetime of a typical U.S. family sedan is 820,000 MJ, about five times more energy than is used in producing the vehicle (Figure 4-11).

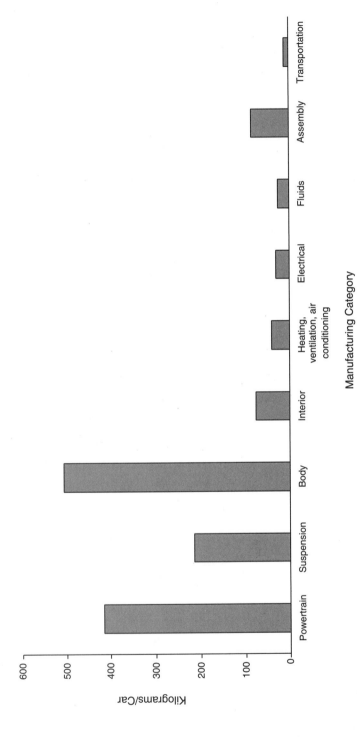

FIGURE 4-7 Solid waste produced during automobile manufacture. SOURCE: Fischer and Schot (1993).

TABLE 4-2  Recycled Waste Types and Quantities at General Motors, 1996

| Waste Type | Pounds per Vehicle Produced | Percent Recycled |
|---|---|---|
| Metal | 600 | 99 |
| Synthetic oils | 0.5 | 99 |
| Batteries | 4.2 | 97 |
| Empty drums | 2.2 | 95 |
| Paper | 34 | 86 |
| Glass | 0.1 | 79 |
| Plastic | 7.4 | 78 |
| Wood | 17 | 73 |
| Paint process organics | 0.7 | 71 |
| Aqueous process liquids | 0.5 | 61 |
| Paint shop maintenance | 0.03 | 60 |
| Petroleum oils | 19 | 58 |
| Paint purge organics | 0.2 | 50 |
| Derived/mixture[a] | — | 5 |
| Other | 112 | — |
| Total | 800 | 60 |

[a]Reportable under RCRA.
SOURCE:  Adapted from General Motors (1997).

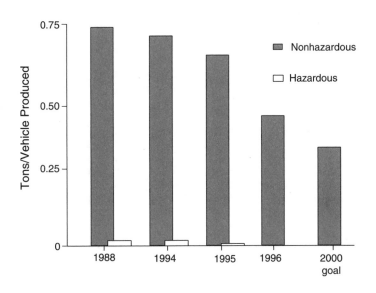

Year

FIGURE 4-8  Hazardous and nonhazardous solid waste per vehicle produced.  SOURCE: General Motors (1997).

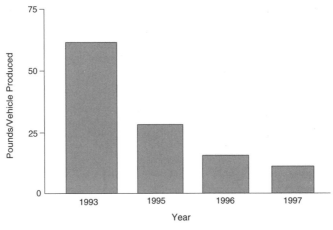

FIGURE 4-9  Packaging waste per vehicle produced.  SOURCE:  General Motors (1998).

---

**BOX 4-4  Plastics Reuse in Manufacturing**

The Chrysler Kokomo Transmission Plant, located in Kokomo, Indiana, manufactures numerous transmission and subassembly components from parts made on-site and obtained from outside sources.  The plant uses more than 20 color plastic caps and plugs during assembly of the transmission to protect critical openings and electrical connections from dirt and contaminants and from damage during shipping.  Some of the caps and plugs are removed during installation at the assembly plant.  The caps and plugs are made of various types of plastic, making them difficult to recycle.

Reuse options for the caps and plugs were reviewed by Kokomo Transmission Plant personnel working with Chrysler's National PQI Waste Elimination Team (a joint union-management product quality improvement (PQI) initiative with the goal of reducing waste).

Two options were proposed:  elimination of the caps and plugs altogether and changing all of the caps and plugs to a uniform color and type of plastic (clear, low-density polyethylene).  The first option provided no flexibility; the second, on the other hand, helped the plant recycle the caps and plugs by reducing the variability and increasing the volume of material.  The change was made in direct consultation with product engineers and by changing procurement requirements for Chrysler suppliers.

Chrysler assembly plants now collect a few different types of caps and plugs and ship them back with the empty transmission racks to the transmission plant.  The transmission plant unloads the caps and plugs and ships them to a local workshop, where they are washed, sorted, and repackaged for reuse at the transmission plant in place of new ones.

In 1996 approximately 50,000 pounds of caps and plugs were reused, and 1.2 million pounds of plastic were recycled.

SOURCE: Michigan Department of Environmental Quality (1998e).

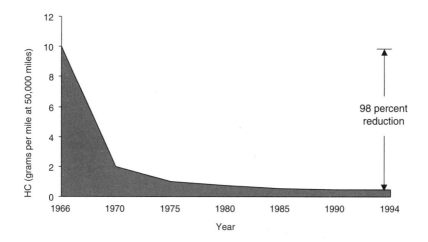

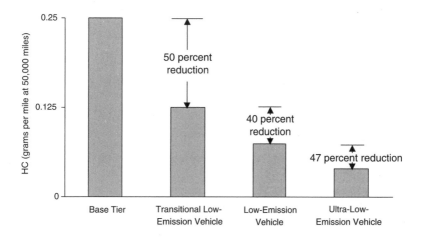

FIGURE 4-10  Evolution of U.S. hydrocarbon (HC) emission standards, 1966–1994 (top), and predicted development of post-1994 low-emission vehicles (bottom). SOURCE:  Ford Motor Company (1997).

## Reuse, Recycle, and Disposal

Vehicle recyclability has received considerable attention recently because of voluntary European take-back initiatives, which are aimed at making manufacturers responsible for the entire life cycle of their products (including final disposal).  In the United States, as mentioned earlier, more than 75 percent of the mass of the average vehicle is put back into useful products.  The processes used to recycle automobiles and automobile materials are shown in Figure 4-12.  The

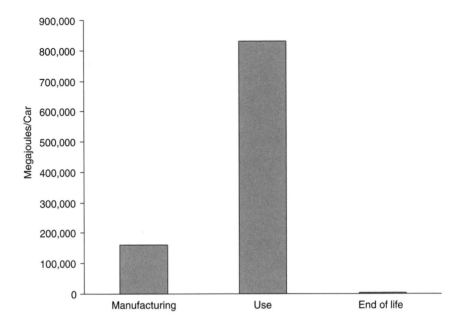

FIGURE 4-11  Total primary energy use over the life of a car.  SOURCE:  U.S. Automotive Materials Partnership, Life-Cycle Assessment Special Topics Group (1997).

economics of recovering these materials depends on the condition of the vehicle being recycled.

While U.S. automobile manufacturers do not directly recycle vehicles, they are involved in a vehicle recycling partnership (VRP).  Established in 1991, VRP's major goals are to reduce the total environmental impact of vehicle disposal, increase the efficiency of disassembly processes to enhance recyclability, develop material selection and design guidelines, and promote socially responsible and economically achievable solutions to vehicle disposal.  While initiatives like these may produce their own metrics, the metric typically used for recycling and recovery is the percentage of a vehicle recycled.  While not used directly in marketing efforts, this metric is one of several that could be used to measure the "eco-friendliness" of vehicles, a theme of the 1998 Japanese, American, and European auto shows.

## Present State of Environmental Metrics

A substantial number of environmental performance metrics are currently used and reported by the automotive industry.  Table 4-3 lists the metrics being

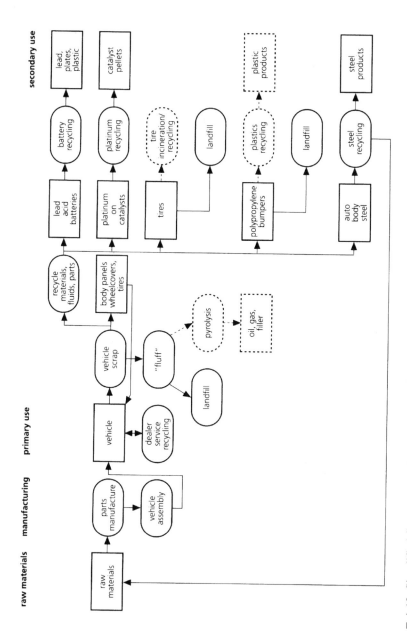

FIGURE 4-12 Simplified flow process diagram for generic automobile materials. SOURCE: France and Thomas (1994).
NOTE: Rectangles indicate products; ovals indicate processes.

TABLE 4-3  Environmental Performance Metrics, Drivers, and Environmental and Business Motivations

| Metric | Driver | Environmental and Business Motivations |
|---|---|---|
| Manufacturing emissions (CM1) | R | Federal and local discharge limits and operating permits |
| Toxic release inventory (CM2) | R | EPA list associated with health concerns |
| 33/50 Chemicals (CM3) | V | 17 selected chemicals from TRI list for early action |
| Great Lakes persistent toxics (CM4) | V | Concern for Great Lakes ecosystems and transfer to humans via food chain |
| SARA Title III chemicals (CM5) | R | Concern for soil and groundwater contamination |
| Energy (E1) | V | Reduce total energy consumption and operating costs |
| Materials (M1, M1R) | V | Improve materials use efficiency |
| Water (M2, M2R) | V | Reduce water use and operating costs |
| Packaging waste (M3) | V | Reduce landfill waste and disposal cost |
| Solid waste (M4) | V | Reduce landfill waste and disposal cost |
| Tailpipe HC and $NO_x$ emissions (PM1) | R | Reduce urban ozone to protect public health |
| Evaporative emissions (PM2) | R | Reduce HC emissions contribution to tropospheric ozone production |
| Tailpipe $CO_2$ emissions (PM3) | R | Limited via CAFE standards to reduce gasoline consumption and global warming gases |
| Recycled waste (PM4) | V | Reduce landfill waste and disposal cost |
| Human health and safety (HHSM) | R | Healthy and safe workplace for employees, safe products for customers |

NOTES: CM1, CM2, CM3, CM4, CM5, E1, M1, M1R, M2, M2R, M3, M4, PM1, PM2, PM3, PM4, HHSM refer to metrics shown in Figure 4-3.  R = regulation driven;  V = voluntary.

used, identifies them as either regulation driven or voluntary, and presents the business or environmental motivation behind the metric.

Automotive industry environmental performance metrics can also be presented according to whether they are used in manufacturing or with end products (i.e., vehicles; Table 4-4). Product metrics can guide decision making in the design and development stages. Human health and safety metrics—both manufacturing and product related—are included in this set of metrics, since they have historically been a key component of environment, health, and safety management programs.

## CHALLENGES AND OPPORTUNITIES

A plethora of environmental performance metrics is currently used internally by the auto industry. Many are driven by regulatory requirements, others by self-interest. The pressure is on—locally, nationally, and globally—to continually improve the environmental performance of the industry's operations and products. In addition, the globalization of the automotive industry and the search for fundamentally new vehicle technologies is expected to challenge the competitiveness of the industry throughout the coming decades. As demand for im-

TABLE 4-4   Environmental Performance Metrics Used in Automotive Manufacturing and Vehicle Use

| Manufacturing | Vehicle Use |
|---|---|
| *Resource Related* | |
| Energy used in manufacturing by facilities (total by company and per vehicle produced) | Gasoline consumed (miles/gallon) |
| | Maintenance materials (per vehicle) |
| | Repair parts (per vehicle) |
| | All metals recycled (75% of vehicle mass) |
| Water used | |
| | |
| *Environmental Burden Related* | |
| TRI | Tailpipe HC, CO, and $NO_x$ emissions (grams/mile) |
| 33/50 Chemicals | Evaporative emissions (grams/test) |
| Great Lakes persistent toxics | Tailpipe $CO_2$ emissions (grams/mile) |
| SARA Title III chemicals | Landfilled shredder residue |
| Solid waste (per vehicle) | (25% of vehicle mass) |
| Gaseous/liquid emissions | |
| | |
| *Human Health and Safety* | |
| Oil mist | Crash worthiness |
| Hazardous chemicals handling | Occupant protection |
| Plant-level noise | Driver behavior |
| Ergonomics | |

proved environmental performance grows, metrics will be needed that are easy to understand, related to quantifiable environmental impacts, and based on good business principles.

## Emerging Issues

The automotive industry produces vehicles that many regard as necessities in the modern world. As a result, the industry's environmental goals are strongly interwoven with those of society and government. Significant engineering challenges lie ahead. In the near term, demands for more fuel efficient vehicles are expected to grow as nations around the world try to reduce $CO_2$ emissions to address concerns about global warming. Meeting the demand for personal mobility on a global scale and addressing the associated traffic congestion problems will pose an additional challenge and add to the complexity of developing relevant environmental performance metrics to guide decision making.

### Continuation of Increased Demand

Vehicle use is ubiquitous. There were approximately 200 million vehicles registered in the United States and an additional 450 million registered in other parts of the world in 1996 (America Automobile Manufacturers Association, 1997). These vehicles consume large amounts of fossil fuel every year, thereby contributing to such environmental burdens as air pollution. The overall energy impact of vehicle use exceeds the energy impact of vehicle manufacture by about a factor of five, a situation similar to that in the electronics sector. (See Chapter 6.) Although the amount of pollution generated per vehicle has decreased due to more efficient pollution control technologies, the demand for vehicles continues to increase worldwide; thus, the amount of pollution generated by vehicle production and use continues to rise.

The large number of vehicles in use and the potential growth in demand for new vehicles leads to a huge aftermarket for used vehicles. This may have various environmental impacts as a consequence of the way such vehicles are operated, serviced, recycled, and disposed of. Regulations on vehicle emissions apply to both new and in-use vehicles. For example, emissions control systems are designed to meet a required durability metric of 100,000 miles. In both the new- and the used-vehicle markets, the industry is served by a large number of small service providers and dealers engaged in recycling parts and spent fluids (e.g., engine and transmission oil, brake fluids, and coolants). Ensuring the adequacy of these practices and the ultimate disposal of used vehicles is a serious concern.

Many of the emissions issues related to vehicle manufacturing are similar to those in other industries. Their management, however, poses special challenges to the automotive industry and others (e.g., the aircraft industry) that deal with a vast supplier base and complex assembly operations. In the case of automobile manufacturing, production involves thousands of small and large suppliers in many manufacturing industries around the world. Each automobile company has a unique collection of parts and processes. The selection of materials and processes can change frequently within a plant and can be different from one plant to the next and from one supplier to another. Deciding on what type of information to track and report (other than that required by regulation) can be a daunting task. When it comes to monitoring environmental performance, data on manufacturing emissions from individual plants tend to give a fragmented view. Aggregated amounts and types of emissions from all plants may provide a more complete picture but very little guidance for taking action. Goals set at the corporate level are important to that prioritization.

## Business Implications

Traditionally, costs associated with handling, treating, and disposing of manufacturing wastes are not fully incorporated as line items in accounting records. This leads to missed opportunities to highlight cost savings and waste reduction achieved through conservation of resources, recycling, and reuse. Efforts to capture some of these hidden costs include Chrysler's LCM evaluation of nontraditional costs as part of the company's search for low-cost alternatives that reduce environmental, health, and safety impacts. In general, however, these techniques still require improvement.

Other costs not fully accounted for include those associated with owning and operating vehicles (e.g., repairing or replacing pollution control devices) and those related to vehicle use such as construction and maintenance of roadway systems. Varied tax and incentive schemes for different types of fuels, as well as different fuel taxes in different parts of the world, impact these and other environmental costs.

There are also questions about how one captures tangible and intangible environmental benefits. Some cost avoidance or savings due to better environmental management (e.g., reuse, recycling, material replacement, waste minimization) can be quantified. However, there are many environmental benefits that are difficult to incorporate into the bottom line. It is also unclear whether there is a sufficient customer base for "environmentally friendly" products or whether customers are willing to pay more for a product's environmental attributes. Evidence from the types of vehicles that people purchase and from company market surveys indicates that customers consider environmental improvements to be important features but unimportant to the final purchase decision. Finally, there are many rules of thumb regarding what constitutes an environmentally preferable manufacturing process (e.g., reduce, reuse, recycle, eliminate), but there is little agreement on key definitions or on how to quantify environmental superiority. These important issues need to be addressed if the environmental performance of a company is to be measured and compared with that of other firms in a meaningful way.

## Globalization

As the automotive industry becomes more global, issues related to manufacturing and vehicle emissions become increasingly complicated. Regulations for controlling and managing emissions vary from one country to another. Economic conditions and public expectations for environmental performance also vary. As a result, environmental management needs to be carried out locally to meet local regulations but with a global perspective.

The industry expects to face new environmental standards. Although regulations have led to some innovations, regulatory approaches do not elicit the most creative solutions. Flexible incentive-based approaches will more effectively stimulate future innovations. Risk-based approaches, which prioritize concerns on the basis of environmental or health risks, are gaining acceptance in the industry as a means of defining environmentally significant emissions. These can also act as a basis for setting standards globally. Currently, companies handle differences in global standards and regulations by harmonizing and standardizing environmental practices across geographical, political, and cultural boundaries. Such steps may include instituting environmental auditing, waste control, treatment facility design, pollution prevention, waste minimization, resource conservation, and risk reduction programs.

## Standardization and Definition

The automotive industry has considerable data on environmental metrics. These can be assembled in various forms. For example, GM's environmental report complies with Coalition for Environmentally Responsible Economies guidelines and provides detailed information about its environmental performance. The absence of similar types of reporting makes comparisons of environmental performance among different corporations difficult. A key challenge for the auto industry is international standardization of environmental performance metrics and reporting practices.

Interpretations of environmental metrics in automotive manufacturing need to take into account the different degrees of vertical integration among U.S.-based companies. (Vertical integration reflects the extent to which a company manufactures the parts it needs for production.) A metric expressed in pounds per vehicle produced may be different for otherwise comparable vehicles because of differences in vertical integration and supplier chains among manufacturers. Metrics need to account for these differences. The same difficulty applies to comparison of financial metrics. Intercompany comparisons aside, however, the numerical values for a specific company are important for tracking its progress over time. Standardizing and defining metrics will be critical to improving environmental performance over the near term.

## Emerging Opportunities

Several foreign-owned and U.S. automobile companies are already using environmental criteria to advertise and market their products. On the investment front, some in the financial community are beginning to use an environmental performance screen as an additional tool in making investment decisions (Deutsch, 1998). As a result, companies of equal standing in all other areas gain an advantage by demonstrating superior environmental performance.

## New and Possible Future Metrics

Many of the metrics currently used are empirical, defined somewhat intuitively, and have their origin in regulations. The automotive industry seeks metrics that are easy to understand, are related to quantifiable environmental impacts, and are tied to business performance metrics such as return on assets, customer responses, and operating costs. If environmental performance metrics can be linked to business performance metrics, corporate efforts to reduce environmental impact will follow.

### Development and Implementation of New Environmental Performance Metrics

Before developing an environmental performance metric, it is important to evaluate how the metric will be used. Considering the complexity of the automotive industry, simple empirical metrics may provide a limited view of overall environmental performance. Misuse of such a metric could be more confusing than not using it at all. For example, a metric such as pollutant generated per vehicle produced could be misleading if it were used to compare plants either within a company or between companies, because each plant may have different starting materials and processes even if each is producing the same end product. Therefore, clear boundaries need to be defined for any comparison and the results of such comparisons carefully evaluated.

Using life cycle inventory (LCI) as a basis for developing environmental metrics is appealing because the intent of the methodology is to consider many environmental burdens in a systematic way. The complete methodology entails doing an inventory of inputs and outputs and an assessment of associated environmental impacts. Despite attempts to incorporate weighted impacts in LCIs (Horkeby, 1997), its primary application is in conducting inventories to guide decision making. The LCI considers all aspects of producing vehicles, starting from materials and parts production (suppliers) to final assembly by an automotive company, use of the vehicles by customers, and end-of-life disposal. The results of an LCI study show the energy profile for producing, using, and disposing of a vehicle, as illustrated in Figure 4-11. In this example the vast majority of energy used by a vehicle over its lifetime is consumed during its service life. This shows just one aspect of the LCI, and the metrics based on energy usage alone could be misleading. Figure 4-8, which presents lifetime water usage data, shows the dominance of manufacturing, quite a different picture from energy consumption.

Another factor in developing viable environmental performance metrics is the differences in market demand and economic growth between industrialized and industrializing countries. Metrics that merely report on total quantities of emissions or resources do not capture the differences in economic development among different countries. Comparisons of environmental performance between companies in different countries have to be made with care, taking into account differences related to such things as gross domestic product, the maturity of the industry, infrastructure, and labor costs.

## Possible Future Metrics

There are several candidates for new metrics that could be used to gauge environmental performance in the automotive industry. None provides a comprehensive picture of the industry's environmental performance. Therefore, their limitations should be carefully considered if they are to be used to make comparisons in the environmental performance of different companies or industries.

### Resource-Use and Pollutant-Generation Metrics

Environmental performance can be linked to the quantities of resources, including energy, raw materials, water, and air, required to produce vehicles. Resource use could be expressed as water ($m^3$)/vehicle, air ($m^3$)/vehicle, energy (kWh)/vehicle, steel (tons)/vehicle, or plastics (tons)/vehicle. A method could be devised to combine some of these. These units would provide a measure of the efficiency of manufacturing processes and product use. These units, however, would reflect environmental impacts only indirectly.

Metrics related to the pollutants generated during vehicle manufacture and use could be expressed as tons of TRI chemicals, hydrocarbons, $CO_2$, toxic metals, and chemical oxygen demand (COD) released per vehicle. These metrics would show the efficiency of controlling the generation of pollutants.

By combining the above two groups of metrics, normalized metrics could be developed to reflect efficiencies of resource use and pollution minimization. These might reflect tons of TRI chemicals released/$m^3$ of water used, tons of TRI chemicals released/kWh used, tons of total hydrocarbons released/ton of steel used, and tons of COD released/$m^3$ of water used.

These normalized metrics would have several potential advantages. They would be easy to understand, could be used for the whole product life cycle or individual manufacturing processes, would reflect the degree of resource recycling and reuse, and would relate to sustainable development.

*Business-Related Metrics*

One glaring deficiency in the current set of environmental performance metrics is the lack of a metric that is strongly linked to customer satisfaction. For example, in the case of CAFE, it has been shown that while energy efficiency improvements to vehicles have been made for a range of different reasons, there is a dissonance between efficiency standards and market signals.[4] Metrics that capture less tangible values, such as a vehicle's "environmental performance," in terms of customer "utility," would marry both environmental and business imperatives. In this circumstance, environmental performance and utility are subjective and need to be further defined. For example, utility could mean just getting from point A to point B, or it could also include deriving a certain level of pleasure from the trip. Such societal metrics would vary from one culture to another. These more sophisticated metrics are currently missing in the suite of simpler resource use and pollutant-related metrics.

There are a range of other measurable business metrics that have environmental implications. For example, the durability of emissions control systems has a huge effect on used-car resale values in many countries. There is a disconnect between metrics such as the frequency of repair or a warranty extension option on emissions control systems and environmental performance metrics. There are two needs in this regard: a better understanding of system interactions,

---

[4]The CAFE approach to improving energy efficiency (often a surrogate for environmental improvement) places manufacturers in an awkward position in the marketplace. According to the National Research Council (1992):

> The [CAFE] system constrains consumer choice—at least in the aggregate—to vehicles with characteristics that differ from those that otherwise would be offered. In effect the manufacturers are required to sell vehicles with higher fuel economy regardless of consumer interest in purchasing such vehicles. A rational consumer—other things being equal—should always prefer a fuel-efficient vehicle over a fuel-inefficient vehicle. But, in fact, an increase in fuel economy can impose costs: the financial costs of the technology to achieve fuel economy and the costs reflected in reduced vehicle size, decreased performance, and other undesirable attributes. To the extent that improved fuel economy is achieved at the expense of other characteristics of the vehicle that the consumer values more, the manufacturer is being required to try to sell a product that does not reflect consumer demand. The consequences are an undesirable diminution in consumer satisfaction and, presumably, a reduction in overall automotive sales. In short, over the last several years, there has been a growing dissonance between CAFE standards and market signals as the real price of gasoline has fallen. (p. 169)

such as the connection between the life of emissions control systems and the life of the entire vehicle, and incentives for developing better modular upgradable systems that can lead to longer warranties.

Table 4-5 summarizes several possible future metrics that may be used to further refine the assessment of the environmental performance of the automotive sector.

## SUMMARY

While many of the metrics associated with the automotive sector are driven by regulation, some are voluntary and business driven. Wherever possible, the latter two types of metrics should be integrated into assessments of environmental performance. For clarity the metrics being used in the automotive industry were discussed in terms of manufacturing and product use. Metrics used for manufacturing include emissions (e.g., TRI releases, VOC emissions, and wastewater discharges), recycling and reuse (e.g., solid waste generation and reuse of plastics), and resource use (e.g., energy and chemicals). Metrics used to monitor product performance include tailpipe emissions (e.g., CO, $NO_x$, and hydrocarbons) and resource consumption (e.g., fossil fuel use and $CO_2$ generation).

The automotive industry has been steadily improving its environmental performance, but many challenges to developing ideal metrics remain. These in-

TABLE 4-5 Potential Future Environmental Performance Metrics, Drivers, and Environmental and Business Motivations in the Automotive Industry

| Metric | Driver | Environmental and Business Motivations |
| --- | --- | --- |
| Resource use | V | A measure of efficiency in manufacturing and product usage |
| Pollutant generation | R | A measure of efficiency in controlling pollutants |
| Pollutants generated per resource used | V | A measure of efficiencies in both resource use and pollutant generation |
| Utility-based environmental performance | V | A measure of environmental performance based on customer utility |
| Repair frequency of pollution control devices | V | Reflects the quality and resale values of vehicles and customer satisfaction |
| Warranty cost on pollution control devices | V | Reflects the quality and resale values of vehicles and customer satisfaction |

NOTE: R = regulation driven; V = voluntary.

clude combining environmental, social, global, and cultural issues associated with producing and using vehicles worldwide; quantifying intangible environmental benefits and incorporating them into the business bottom line; taking into account differences and priorities across countries, particularly with regard to industrializing and industrialized countries; developing weighted priorities based on environmental risk; and balancing the demand for simple robust metrics with an appreciation of their interpretive limitations.

## REFERENCES

American Automobile Manufacturers Association (AAMA). 1994. AAMA Motor Vehicle Facts and Figures. Detroit: AAMA.

American Automobile Manufacturers Association (AAMA). 1997. Motor Vehicle Facts and Figures. Detroit: AAMA.

American Automobile Manufacturers Association (AAMA). 1998. Environmental Responsibility: Progress Measurement. Available online at http://www.aama.com/environment/progress.html. [Jan. 13, 1999]

Deutsch, C. 1998. For Wall Street, increasing evidence that green begets green. New York Times. July 19, Section 3, p. 7.

Fischer, K., and J. Schot. 1993. Environmental Strategies for Industry: International Perspectives on Research Needs and Policy Implications. Washington, D.C.: Island Press.

Ford Motor Company. 1997. Environmental Annual Report, Ford Motor Company. Detroit: Ford Motor Company.

Ford Motor Company. 1998. Design for Environment. Available online at http://www.ford.com/corporate-info/environment/research/design4enviro.html. [August 11, 1998]

France, W., and V. Thomas. 1994. Industrial ecology in the manufacturing of consumer products. Pp. 339–348 in Industrial Ecology and Global Change, R. Socolow, C. Andrews, F. Berkhout, and V. Thomas, eds. London: Cambridge University Press.

General Motors. 1997. The Right Road. General Motors Environmental Health and Safety Report. Detroit: General Motors Company.

Graedel, T.E., and B.R. Allenby. 1997. Industrial Ecology and the Automobile. Upper Saddle River, N.J.: Prentice Hall.

Horkeby, I. 1997. Environmental prioritization. Pp. 124–131 in The Industrial Green Game, D.J. Richards, ed. Washington, D.C.: National Academy Press.

Kainz, R.J., M.H. Prokopyshen, and S.A.Yester. 1996. Life cycle management at Chrysler. Pollution Prevention Review 6:71–83.

Lee, R., M. Prokopyshen, and S. Farrington. 1997. Life cycle management case study of an instrument panel. Paper No. 971158 in the Proceedings of the Total Life Cycle Conference, Society of Automotive Engineers, April 6.

Michigan Department of Environmental Quality. 1998a. Pollution Prevention in the Auto Industry. Available online at http://www.deq.state.mi.state/p2sect/auto. [August 11, 1998]

Michigan Department of Environmental Quality. 1998b. Process Change to Eliminate the Use of Trichloroethylene. Available online at http://www.deq.state.mi.us/ead/fact/auto/ford.html#PROCESS CHANGE TO ELIMINATE THE USE OF. [August 11, 1998]

Michigan Department of Environmental Quality. 1998c. Chrysler Corporation Underhood Mercury Switch Life Cycle Management Study. Available online at http://www.deq.state.mi.us/ead/fact/auto/chrysler.html. [August 11, 1998]

Michigan Department of Environmental Quality. 1998d. Turning Off the Water Saves Millions of Gallons. Available online at http://www.deq.state.mi.us/ead/fact/auto/gm.html. [August 11, 1998]

Michigan Department of Environmental Quality. 1998e. Kokomo Transmission Plan Reuses Plastic. Available online at http://www.deq.state.mi.us/ead/fact/auto/chrysler.html. [August 11, 1998]

National Research Council. 1992. Automotive Fuel Economy: How Far Should We Go? Washington, D.C.: National Academy Press.

United States Automotive Materials Partnership, Life-Cycle Assessment Special Topics Group. 1997. Life-Cycle Inventory Analysis of a Generic Vehicle. Detroit: Automobile Industry SubGroup.

United States Department of Commerce (USDOC). 1995. Inventions Needed for PNGV. Washington, D.C.: USDOC.

United States Environmental Protection Agency (USEPA). 1999. Superfund: Cleaning Up the Nation's Hazardous Waste Sites. Available online at http://www.epa.gov/superfund/index.htm. [February 12, 1999]

# 5

# The Chemical Industry

## BACKGROUND

The chemical industry is more diverse than virtually any other industry in the United States. Harnessing basic ingredients, the industry[1] produces a plethora of products not usually seen or used by consumers but that are essential components of, or are required to manufacture, practically every consumer and industrial product (Box 5-1). Many chemical industry products are intermediates, and chemical company customers are often other chemical companies. Several companies in this industry are also at the forefront of emerging biotechnology industries.

According to the American Chemical Society (1998), the industry's more than 70,000 different registered chemical products are developed, manufactured, and marketed by more than 9,000 companies.[2] Of these firms, 40 account for roughly half the industry's output on a mass basis. The chemical industry is the third-largest manufacturing sector in the nation, representing approximately 10 percent of all U.S. manufacturing and boasting one of the largest trade surpluses

---

[1]Eight standard industrial classification codes are used by the U.S. Department of Commerce to categorize chemical companies. The categories are industrial inorganic chemicals; plastics, materials, and synthetics; drugs; soap, cleaners, and toilet goods; paints and allied products; industrial organic chemicals; agricultural chemicals; and miscellaneous chemical products.

[2]Total revenues of the U.S. chemical industry in 1997 were about $400 billion, or 1.9 percent of total gross domestic product. Capital investment is approximately $35 billion a year, and research and development spending is approximately $18.3 billion per year (Chemical Manufacturers Association, 1995).

BOX 5-1
Major Raw Materials, Products, and Product End Uses of the
Chemical Industry

*Raw Materials*

| | |
|---|---|
| Air | Oil |
| Coal | Wood |
| Energy | Sulfur |
| Minerals | Seawater |
| Natural gas | |

*Products*

| | |
|---|---|
| Acids | Nylon |
| Alcohols | Pigments/dyes |
| Benzene | Polyester |
| Caustic soda | Polyethylene |
| Esters | Polypropylene |
| Ethers | Polyvinylchloride |
| Ethylene | Solvents |
| Fibers | Synthetic rubber |
| | Xylene |

*Product End Uses*

| | | |
|---|---|---|
| Adhesives | Food ingredients | Pharmaceuticals |
| Automobiles | Fuel additives | Piping |
| Boats | Household materials | Preservatives |
| Carpets | Insulation | Roofing |
| Computers | Packaging | Safety glass |
| Construction materials | Paint and coatings | Soaps and detergents |
| Containers | Paper | Sports equipment |
| Cosmetics | Personal care | Textiles |
| Fertilizers | Pesticides | Tires |
| | | Toys |

of any industry sector ($20.4 billion in 1995). It also ranks as the largest manufacturing sector in terms of production and sales[3] and employs about 1 million people, roughly the same number as the automotive sector.

---

[3]The United States is the world's largest producer of chemicals. The $367 billion worth of U.S. chemical products produced in 1995 represented about 24 percent of the worldwide market, which was then valued at $1.3 trillion. Countries that rank next in production are Japan, Germany, and France. In terms of exports, Germany is currently the global leader. The United States ranks second, capturing approximately 14 percent of total exports worldwide (American Chemical Society, 1998).

In the case of the chemical industry, the committee focused its examination of environmental metrics on the production of commodity and specialty chemicals and polymers. Excluded in the analysis are high-value materials such as cosmetics, food additives, and health care products, including pharmaceuticals, which account for a significant fraction of the industry's sales. The energy and materials demands and waste generation associated with these high-value materials are considerably smaller than those for commodity and specialty chemicals. However, the metrics discussed in this chapter would also be relevant to production of the high-value materials. Another segment of the industry not covered in this chapter is the emerging area of biochemical processing, which may require very different environmental performance metrics.

## Chemical Production

Chemical products result from chemical processes, which are a complex combination of reaction, distillation, absorption, filtration, extraction, drying, and screening operations. For cost-cutting purposes, most chemical processes must be efficient, and so the design of many production operations is focused on controlling and reducing losses of precious materials. Hence, ecoefficiency, including avoiding releases to land, water, or air, is critical to the industry's economic survival. While every chemical process is unique, most can be generalized to a flow diagram, as shown in Figure 5-1.

## Drivers of Improved Environmental Performance

The common driver of environmental performance in the chemical industry, like the automotive industry, is regulation. Public and community concerns about the performance of chemical manufacturers (particularly following negative publicity associated with contaminated waste sites such as Love Canal or chemical accidents such as the tragedy in Bhopal, India) led the industry to establish a set of minimum environmental performance standards.[4] For some companies the costs of complying with environmental regulations are equal to what they spend on research and development.[5] Experience has shown that it

---

[4]The Chemical Manufacturers Association (CMA) launched the Responsible Care program in 1988 in response to public concerns about chemical safety. The program commits CMA member companies to continuously improve health and safety by implementing six codes of management practice: community awareness and emergency response, pollution prevention, process safety, distribution, employee health and safety, and product stewardship. A centerpiece of Responsible Care is the establishment of community advisory panels in locales where member companies have operations.

[5]During the 1980s, DuPont's outlays for regulatory compliance were increasing at about 7 percent per year and totaled over $1 billion per year, which was about equal to the company's entire research and development budget at that time (Carberry, forthcoming).

88

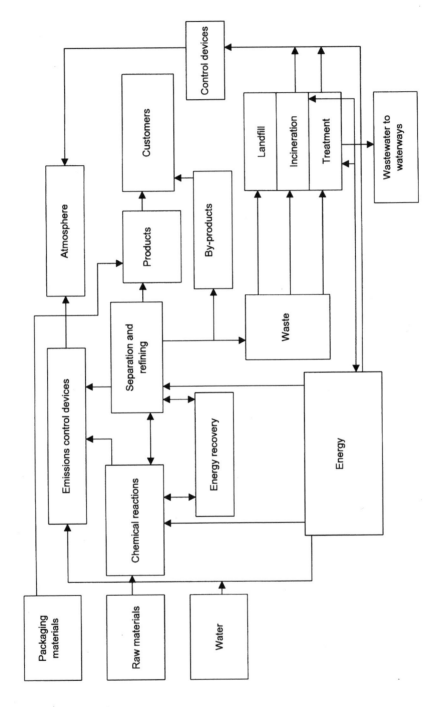

FIGURE 5-1 Schematic diagram of typical chemical manufacturing process.

costs more to react to a regulation than to design systems that address environmental issues from the start (Carberry, forthcoming). For related reasons, some recent environmental performance improvements in the chemical industry have been motivated by the desire to gain a competitive advantage. "Good neighbor" motivations and competitiveness concerns have encouraged some of the larger chemical companies to commit to the (thermodynamically unrealizable) goal of reducing emissions and chemical residuals to zero and to work to eliminate the dispersion of chemicals that adversely impact human health or the environment.[6] More recently, chemical companies (particularly those with plans to grow significantly using biotechnology as a base) have begun to explore using sustainable development as a means of driving change and enhancing environmental performance.[7]

## CURRENT USE OF ENVIRONMENTAL METRICS

Environmental metrics in the chemical industry fall into two broad categories: those related to process efficiencies, such as yield, and those related to product stewardship. These are summarized in the flow chart shown in Figure 5-2. However, to facilitate comparison with the other industries analyzed in this report, the chemical industry metrics are discussed in terms of manufacturing process and product performance.

### Manufacturing Metrics

As in the automotive sector, environmental performance in the chemical industry is monitored and guided by environmental staff, but ultimate responsibility lies with the site operations manager. However, because of the industry's commitment in 1988 to a set of basic environmental standards (Box 5-2), many companies have driven a sense of environmental responsibility throughout their corporate structures. Within the industry's manufacturing operations, this has resulted in an ongoing effort to incorporate metrics into decision making and to continually look for ways to refine current metrics and their uses.

---

[6]Companies like DuPont have adopted the goals of zero waste and zero emissions to drive the reuse and recycling of materials to minimize the need for treatment or disposal and to conserve resources. The drive toward zero emissions gives priority to those emissions that pose the greatest potential risk to health or the environment. In some cases this zero discharge approach has led to the development of new processes that eliminate the need for certain chemicals as raw materials or as by-products. In others it has led to changes in management practices.

[7]For example, Monsanto spun off Solutia as a traditional chemical company to focus on biotechnology. Monsanto has made sustainable development the cornerstone of its environmental policy and has begun to develop goals and metrics in line with several sustainability principles (Magretta, 1997). The policy has helped counterbalance public concerns about Monsanto's efforts to create genetically modified crops for improved food production.

90

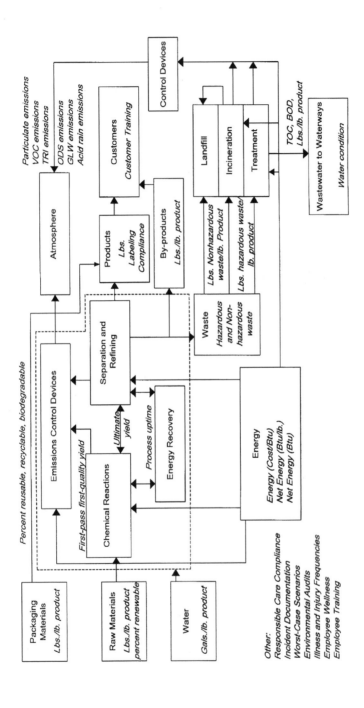

FIGURE 5-2  Metrics used in chemical manufacturing.
NOTE: VOC = volatile organic compound, TRI = toxic release inventory, ODS = ozone depleting substances, GLW = Great Lakes waste, TOC = total organic carbon, BOD = biological oxygen demand.

BOX 5-2
Chemical Manufacturers Association
Responsible Care Program

In 1988 the Chemical Manufacturers Association (CMA) launched Responsible
Care in response to public concerns about the manufacture and use of chemicals.
Through Responsible Care, member companies support efforts to improve the
industry's responsible management of chemicals. Responsible Care is an obliga-
tion of membership in CMA and requires member companies to

1) improve performance in health, safety, and environmental quality
   • The Guiding Principles—the philosophy of Responsible Care—outline each
     member company's commitment to environmental, health, and safety
     responsibility in managing chemicals.
   • Six Codes of Management Practices, which are the heart of Responsible
     Care, outline practices that cover virtually every aspect of chemical manu-
     facture, transportation, and handling.

2) listen and respond to public concerns
   • CMA's National Public Advisory Panel, a group of leaders in the environ-
     mental, health, and safety fields, provides the public perspective on Re-
     sponsible Care and its activities.
   • The Public Outreach Program targets key audiences and develops means
     for helping member companies open dialogues with them.

3) assist each other to achieve optimum performance
   • Executive leadership groups bring key executives from member companies
     together regularly to share experiences and review progress in Responsible
     Care.
   • The Mutual Assistance Network focuses on direct networking between
     companies at different levels to assist in implementation of all elements of
     Responsible Care.

4) report their progress to the public
   • Member self-evaluations provide a measure of member company progress
     and are a valuable management tool.
   • Performance measures are being developed or exist already that indicate
     the progress CMA member companies are making in carrying out Respon-
     sible Care.
   • Management systems verification will assist member companies improve
     by including appropriate third-party involvement.

SOURCE: Chemical Manufacturers Association (1998a).

**Emissions**

As with the automotive sector, one of the chemical industry's primary environmental concerns is emissions from its manufacturing operations. These are regulated by federal, state, and local laws. Chemicals regulated under the Environmental Protection Agency's (EPA's) Toxic Release Inventory (TRI) and other elements of Title III of the Superfund Amendments and Reauthorization Act of 1996 (SARA) are measured by total weight. Although total weight is appropriate for measuring continuous improvement, it does not take into account different production levels at various facilities. This metric would be more useful if expressed as emissions per mass of product. However, normalizing emissions in terms of mass of product would hide the impact of vertical integration (a key factor in chemical manufacturing) and would fail to account for the value of the product to society. As refinements to emissions metrics and other environmental performance metrics are made, it would be worth considering normalization against dollar value of product or, better still, dollar of value added to society. The latter is difficult to determine but would perhaps be more appropriate for sustainable-development discussions.

The chemical industry has developed a variety of metrics relating to "reportable releases." Through TRI, SARA requires companies to report on a wide range of chemicals, if minimum release quantities are exceeded.[8] In addition to this mandatory reporting, another part of SARA[9] provides for a wide range of voluntary data collection.

Required reports such as TRI are used in the industry to learn about the performance of competitors. Equally important, TRI can be used to identify chemical companies that are having problems with their emissions and which may therefore be in the market for innovative products or approaches that reduce them. Reportable release data can be very confusing in the absence of additional analysis. Several companies, including Dow, DuPont, Monsanto, and Union Carbide, have developed simplified scoring systems that divide this reporting into three categories of significance related to human health. Metrics on all three categories are used internally within the companies for decision-making purposes. The most significant category is reported publicly as incidents (of releases) per year.

Another improvement in emissions metrics relates to predicting toxic dispersion or the potential for toxic dispersion. In the case of airborne emissions, some in the chemical industry are using models to predict the concentrations of pollutants that may occur over a given distance or area. Models are also used to evaluate the reliability, quality, or risk associated with a supplier or as a factor in assessing risks related to potential mergers or acquisitions. In addition, they are

---

[8]Specifically, Section 302 of the Emergency Planning and Community Right-to-Know Act.
[9]Ibid., Section 202.

used for long-term technology or business planning, particularly to prevent the environmental release of materials that pose the greatest risk in terms of toxicity and dispersion potential.

Ideally, emissions metrics should allow one to rank environmental and health risks. This would require a weighting system that takes into account the effects of a range of factors, such as persistent bioaccumulative toxicity, ozone depletion, global warming, atmospheric and surface water acidification, human health effects, photochemical ozone generation, aquatic oxygen demand, and aquatic toxicity.[10] Current regulatory-derived metrics treat all emissions alike and do not take into account differences in hazard potential. One current effort to address this situation is a system of potency metrics developed by Imperial Chemical Industries, a U.K. chemical company (Box 5-3). While a significant step forward, this method can be further improved if modified to weight various chemicals with regard to method of disposal (air, water, landfills, deep-well injection, public treatment facilities, or incineration) and their health exposure "pathway" (i.e., how a particular emission may reach a susceptible receptor in an organism).

## Public Concerns

The potential for a catastrophic chemical release is a major concern of the chemical industry. While there are too few such events for metrics to be applicable, both environmental laws and the Occupational Safety and Health Act require defining a "worst case" and a "more likely" release scenario for a catastrophe and reporting that to the community.

Chemical plumes, incinerators, noise, landfills, regulated outfalls, remediation sites, and transportation accidents can create concern among the public. Community complaints and violations of state or federal regulations are tracked and dealt with by external affairs groups and community advisory panels according to CMA guidelines. Some sites try to anticipate concerns by conducting public opinion surveys about "environmental quality." The results of such surveys are used for community relations as well as for internal planning purposes.

## Resource Use and Waste

Two types of materials are generally tracked throughout the chemical industry: those intended for inclusion in the product ("raw materials") and all other materials purchased, including "support materials" (e.g., acids, bases, solvents), maintenance materials, and packaging. In this context the waste ratio (Box 5-4),

---

[10]These are categories used by Imperial Chemical Industries, a U.K. chemical company. White and Zinkl (1997) provide a more complete listing and summary of possible criteria for such weightings.

BOX 5-3
Potency Metrics Used by Imperial Chemical Industries

The Imperial Chemical Industries' (ICI) Environmental Burden System ranks the potential environmental impact of different emissions by

- identifying a set of recognized global environmental impact categories on which various emissions to air and water may exert an effect, such as acidity, global warming, human health effects, ozone depletion, creation of photochemical ozone (smog), aquatic oxygen demand, and ecotoxicity to aquatic life;

- assigning a factor to each emission that reflects the potency of its possible impact; and

- multiplying the weight of each substance emitted by its potency factor to determine environmental burden in each impact category.

In assessing potential harm the environmental burden system cannot be used to establish the impact of wastes sent to landfills. This is in keeping with the company's philosophy that nothing should be landfilled unless it is safe to do so.

Other key elements in the environmental-burden approach are that

- individual chemicals can be assigned to more than one environmental impact category;
- each chemical has a specific potency factor for each category, and these factors differ;
- each category has its own characteristics and units of measure;
- burdens for each category cannot be added together to give the total environmental burden;
- environmental burden assumes that all individual operations comply with local regulations; and
- environmental burden does not address local issues such as noise and odor.

SOURCE: Imperial Chemical Industries (1998).

first publicized by 3M, has been used to demonstrate the effectiveness of waste reduction activities and is the most cited nonregulatory-driven environmental metric. A variation on this metric is the material efficiency ratio, or the amount of product sold divided by the amount of all materials purchased, sometimes including packaging (Box 5-4). Waste and other losses are included in the ratio to the extent they are known. Some who prefer this latter metric feel that tracking

BOX 5-4
Waste and Material Efficiency Ratios

The waste ratio was developed by 3M to encourage conversion of wastes into by-products (residuals that are productively used in manufacturing) and the reduction of waste.

$$\text{Waste Ratio} = \frac{\text{Waste}}{\text{Product} + \text{By-product} + \text{Waste}} = \frac{\text{Waste}}{\text{Total Output}}$$

Because waste is considered to be a sign of inefficient production, the ratio provides an indicator of waste generation as well as product and materials loss. Some, however, prefer the material efficiency ratio over the waste ratio because of the lack of agreement about a definition of "waste."

$$\text{Material Efficiency Ratio} = \frac{\text{Product Sold}}{\text{All Material Purchases}}$$

only waste fails to account for materials burned for energy recovery; destroyed in incinerators, flares, or biological waste treatment units; or otherwise disposed of (sometimes categorized as "nonproduct" output). Resolving the lack of clarity about the definition of a waste might be a useful refinement of metrics such as these but could prove difficult and costly in a materials-intensive sector like the chemical industry.

**Materials Intensity**

Materials intensity metrics are generally the most useful for decision making in the chemical industry. Yield, first-pass first-quality (FPFQ) yield, process "uptime," and waste per mass of product are monitored daily to ensure consistency of operations. Troubleshooting is called for if there are changes in these measurements' short-term averages (i.e., days to weeks). More serious actions, such as improvement of process equipment and process control, can be triggered by changes in the intermediate-term (i.e., weeks to months) averages of these measures. Stable long-term averages, along with atomic (stoichiometric) and materials efficiency, are factors in designing new technology.

Yield is the ratio of the amount of product sold to the amount of product that should have been produced for sale based on the purchase of raw materials and assuming no waste, no side reactions ("perfect" control of chemistry), and no other losses. It is usually expressed as a percent. Yield data can signal the need for a variety of specific actions (Box 5-5).

BOX 5-5
Actions That May Be Guided by Tracking Yield

A technical engineer might analyze the reasons that yield is less than 100 percent and allocate "blame" to equipment problems, operating problems, and basic weaknesses of the production process. This information could then be used by

- operators and mechanics to identify equipment requiring more careful supervision and more preventative maintenance;
- plant engineers to work with vendors to develop more reliable equipment;
- plant technical engineers to institute better training for operators and develop technology for revised operating procedures, conditions, or materials;
- technical specialists to develop and implement better process instrumentation and process control methodology; and
- researchers to develop improved understanding of the existing chemistry or entirely new process steps, including new catalysts, different reactions, and better separations.

FPFQ yield describes the highest-quality product possible without resorting to any capture and recycle of potentially valuable preproduct. Yield is the single most important indicator of environmental efficiency, while FPFQ yield is the dominant metric for effective use of manufacturing capital equipment.[11]

Interestingly, process uptime, usually regarded as a nonenvironmental metric, is frequently the second most important indicator of materials efficiency. Uptime is the percentage of a year that the equipment is operating at intended rates. When equipment is down for maintenance or operating at a slower rate due to production scheduling or lack of demand, there is no increase in waste generated. However, process uptime is a critical environmental metric because waste generation rates during start-ups and shutdowns are frequently far greater than during normal operations.

A newly emerging concept is atomic efficiency, which is the ratio of the output atoms to input atoms based on chemical stoichiometry. It is usually expressed as a percent. While atomic efficiency is usually not important for products derived from a few simple reactions, it is important for more complex chemicals such as drugs and advanced agrochemicals that rely on complicated multiple-step reactions.

---

[11]A process that increases yield from 85 to 95 percent boosts productivity and decreases costs by almost 12 percent. It also results in an almost 67 percent decrease in waste per pound of product.

## Energy Use

As with other industries, in the chemical sector energy efficiency is used to track environmental performance. The chemical industry's energy requirements—like those of most "heavy" industries—are dominated by production operations. Some companies in the industry are also tracking and reporting emissions of greenhouse gases, particularly carbon dioxide ($CO_2$), since such emissions relate to energy efficiency. In the chemical industry, $CO_2$ emissions result directly from fossil-fuel use in heaters, boilers, and other devices or indirectly as a result of purchased energy (e.g., steam and electricity) derived from fossil fuels.

Some companies report emissions of greenhouse gases in mass of $CO_2$ equivalents per year.[12] Increasingly, these emissions are reported normalized per pound of product or per unit product. Internally, companies generally track energy consumption in BTUs per unit process (such as distillation) or per pound of product. When it is a major expense, the energy consumption of a subprocess (e.g., purification) will also frequently be monitored. Most companies will include recovered energy from wastes (e.g., waste solvent and trash diverted to steam incineration) in their energy-use calculations. This use of waste feedstocks as a fuel usually returns less energy per pound of $CO_2$ emitted, but it is far preferable (and more efficient) to incineration or otherwise disposing of the waste without recovering energy. With the increased focus on energy consumption following the Kyoto Protocol and the probable economic disincentives for generating $CO_2$, comparing alternative processes or products in terms of $CO_2$ production may have great utility in the future.

While energy-use metrics are not currently as important to the chemical industry as are materials-use metrics, stable long-term energy averages are traced and used as a basis for process equipment and control improvements; for planning of fundamental new technology; and for developing cogeneration energy partnerships.

## Product Metrics

Unlike vehicles, commodity and specialty chemicals and polymers are often ingredients in final products, rather than end products themselves. Environmental impacts related to these chemicals, therefore, must be considered within the context in which they are transported, used, and disposed of.

---

[12]Other potential greenhouse gases that may be tracked include methane, other volatile organic chemicals, nitrogen oxides, chlorofluorocarbons, and other longer-life volatile compounds that can have greater climate change potential than $CO_2$ per unit of mass (although compared with $CO_2$, they are present in the atmosphere in much smaller quantities).

**Product Stewardship**

The primary objective of product stewardship is to help minimize the safety risks and environmental impacts of the transportation of chemicals and their subsequent processing by customers. Product stewardship also entails designing the product (and associated delivery systems) to minimize adverse safety and environmental impacts during final use and disposal. Specific efforts include staff training in the safe handling of acids and toxic gases; environmental audits and planning with customers to reduce emissions; reformulating chemicals to avoid volatile solvents or ozone-depleting substances such as chlorofluorocarbons (CFCs); developing end-use products with greatly reduced persistence, bioaccumulation potential, or toxicity; providing a mechanism to reuse or recycle a product at the end of its useful life; and investigating the potential for producing biodegradable chemicals and developing "green" suppliers. A major trend within the chemical industry is extending the stewardship concept not only to suppliers, but also to customers. More attention is also being given to the ultimate fate of the product.

The handling of hazardous materials and chemicals by chemical companies or their customers is another aspect of product stewardship. SARA Title III requires that any potential for an explosion or toxic release has to be reported on the basis of "practical worst-case scenarios." The management challenge is to work with the community to ensure that the low probability of such an event is understood.

The burden of regulatory reporting requirements and the added management concerns inherent in the use of hazardous chemicals has led to innovations in the production and delivery of such compounds. Examples include the development of centralized, zero-storage and zero-transportation manufacturing techniques for highly hazardous material (such as phosgene); the manufacture of less-hazardous, next-step intermediates before storage and transportation; and the provision of facilities at the customer's site for point-of-use, just-in-time generation of highly hazardous materials.

No short-term metrics have emerged for product stewardship, other than the degree of implementation of sound management processes. Review of management processes is part of environmental auditing and is usually done annually for planning and goal setting. CMA has developed a set of guidelines in this area (Box 5-6). The guidelines are intended to prompt action when they are reviewed within the highly specific context of an individual business.

**Packaging**

Although many chemical companies consider packaging a separate issue from materials use, packaging is being increasingly recognized as a major, and in some cases dominant, aspect of materials-use metrics. The common packaging metric is total mass of packaging, sometimes normalized per unit of material

BOX 5-6
CMA Product Stewardship Guidelines

The purpose of the CMA Product Stewardship Code of Management Practices is to make health, safety, and environmental protection an integral part of designing, manufacturing, marketing, distributing, using, recycling, and disposing of chemical products. The code provides guidance as well as a means to measure continuous improvement in the practice of product stewardship.

As part of efforts to make health, safety, and environmental considerations a priority in planning for all existing and new products and processes, the CMA code suggests that member companies

- establish and maintain information on health, safety, and environmental hazards and reasonably foreseeing exposures from new and existing products;
- characterize new and existing products with respect to their risk using information about health, safety, and environmental hazards and reasonably foreseeable exposures, as well as establish guidelines for reevaluating current standards or metrics;
- establish a system to identify, document, and implement health, safety, and environmental risk management actions appropriate to the product risk;
- establish and maintain a system that makes health, safety, and environmental impacts—including the use of energy and natural resources—key considerations in designing, developing, and improving products and processes;
- educate and train employees, based on job function, on the proper handling, recycling, use, and disposal of products and known product uses as well as implement a system that encourages employees to provide feedback on new uses, identified misuses, or adverse effects for use in product risk characterization;
- select contract manufacturers who employ appropriate practices for health, safety, and environmental protection or work with contract manufacturers to help them implement proper handling, use, recycling, and disposal practices;
- require suppliers to provide appropriate health, safety, and environmental information and guidance on their products;
- provide health, safety, and environmental information to distributors; and
- provide health, safety, and environmental information to consumers.

SOURCE: Chemical Manufacturers Association (1998b).

---

**BOX 5-7**
**Packaging Innovations Resulting from Product Stewardship**

Examples of recent innovations in packaging include:

- Polymer compounding pellets packaged in a bag of roughly the same material. This allows the customer to use the pellets while they are in the bag, avoids the need to dispose of the bag, and eliminates dust.
- Water-soluble packaging for herbicides. This allows farmers to add the herbicide with its wrapper in a spray-mix tank, thus avoiding the need to dispose of the packaging material that would have been in contact with the herbicides and subject to disposal regulations. This innovation also advances "inherent safety" by reducing worker exposure to chemicals.

---

sold. The metric is heavily used internally and, despite concerns about protecting confidential business information, is shared externally with increasing frequency. The major needs seem to relate to making more-sophisticated containers or other large assemblies and lightweight packaging and developing recycling networks. Generally, packaging that is intended to be waste (e.g., for consumer foods) or that unavoidably ends up in the environment (e.g., fast-food wrappings) would inflict less environmental harm if it were truly degradable. Conversely, it may be preferable to recycle packaging intended for more durable use. Understanding these differences, designing packaging products, and developing appropriate metrics remain challenges for the industry. In the short term, chemical companies have set goals for reducing the amount of packaging that enters or leaves their facilities. In the longer term, companies are working with their customers to develop innovations to reduce packaging and other risks associated with the purchases of chemical products (Box 5-7).

## SUMMARY OF ENVIRONMENTAL METRICS
## IN THE CHEMICAL SECTOR

The metrics used in the chemical sector, as in the automotive sector, can be categorized in terms of resource use and environmental burden. These are shown in Table 5-1, along with health and safety metrics that are generally collected by the same environmental health and safety staff.

## CHALLENGES AND OPPORTUNITIES

The use of environmental performance metrics and improved environmental stewardship techniques within the chemical industry is increasing. While many metrics have been driven by regulation and a general desire to be a "good neigh-

TABLE 5-1   Environmental Performance Metrics in Chemical Manufacturing and for Chemical Products

| Manufacturing | Product Use |
|---|---|
| *Resource Related* | |
| Material intensity<br>• Percent first-pass yield<br>• Percent ultimate yield<br>• Percent process uptime<br>• Percent *atomic efficiency*<br>• Percent *postconsumer waste used*<br>• Material efficiency (unit consumptions, including water/pound of product) | Material intensity[a]<br>• *Value per pound*<br>• *Pounds replaced*<br>• *Resources saved* |
| Energy intensity<br>• BTUs/pound<br>• Total energy use<br>• *Minimum "practical" energy use* | Energy intensity[a]<br>• *Value/BTU used*<br>• *Energy saved by use* |
| Packaging<br>• Total pounds<br>• Pounds/pounds of product | Renewable<br>• *Percent of product*<br>• *Recyclable* |
| *Environmental-Burden Related* | |
| Environmental incidents<br>• Frequency<br>• Severity<br>• Practical worst-case scenario | Packaging<br>• *Recyclable*<br>• *Biodegradable* |
| Toxic dispersion<br>• Airborne toxics<br>• Carcinogens<br>• Volatile organics<br>• Particulates<br>• Acid gases<br>• "Hazardous" wastes<br>• Aquatic toxicity/oxygen demand<br>• Listed hazardous air (and water) pollutants<br>• TRI chemicals (EPCRA Title III Section 313)<br>• 33/50 chemicals<br>.<br>.<br>.<br>etc. | Toxic dispersion<br>• Global warming<br>• Ozone depletion<br>• *Persistence*<br>• *Bioaccumulative*<br>• *Hormone mimics*<br>.<br>.<br>.<br>*etc.* |

*continued*

TABLE 5-1    *Continued*

| Manufacturing | Product Use |
|---|---|
| Product stewardship<br> • Responsible Care<br> • Environmental audits | Product stewardship<br> • Responsible Care |
| Illnesses and injuries<br> • Illness frequency<br> • Injury frequency<br> • *Employee "wellness"* | Product stewardship<br> • Use warnings<br> • User training |
| Hazardous materials handling<br> • Worker training | |

[a]Most product-use-related material and energy intensity metrics deal with the product itself (e.g., value or energy use per pound). These metrics fail to capture the savings in energy or materials that may accrue from the use of the product. The latter frequently far outweighs the product's own materials or energy intensity. For example, plastics that reduce energy and material consumption by enabling lightweighting of cars through substitution for metals have an energy and material profile of their own, which at a large-systems level may be minuscule when compared with their environmental benefits.

NOTE: Italics indicate terms for which there are no agreed-upon definitions. Potential metrics in these areas will depend on developing common definitions and agreement on their scientific underpinnings.

bor," others have been driven by competitive self-interest. The metrics currently in use, however, have several shortcomings, and there are emerging issues such as sustainability that will also require improved metrics.

## Stewardship of Hazardous Materials

Stewardship of hazardous materials is a vital part of the chemical industry's product stewardship efforts. The consequences of inattention to this issue can destroy a company's credibility and ability to operate. Some companies, therefore, keep track of such measures as storage time or distance transported. Currently, all hazardous materials are treated alike. Correcting for risk factors (e.g., true toxicity, exposure pathway, method of transportation, method of handling) could provide valuable information that would help improve management. Understanding of these types of risk analysis, however, is still in the embryonic stage.

## Emissions and Toxics Dispersion Metrics

The chemical industry's efforts to manage emissions from a multimedia perspective (i.e., to manage emissions to all media [water, air, and land] rather one medium at a time) has demonstrated the advantages of such an approach for environmental management (Solomon, 1993). That approach however, does not alter the need to track emissions by specific media. Rather the challenge lies in the development of potency-related metrics. The development of these metrics by ICI (Box 5-3) has renewed interest in weighted and aggregated metrics. Presently, however, no proposed system effectively aggregates the various categories of concern. More important, there is no consensus on what would constitute the most significant categories. The controversy is illustrated by a recent study that aggregated environmental metrics based solely on reported wastes, spills, and enforcement actions, normalized according to sales dollars (Kiernan and Levinson, 1997). At least two companies that had achieved excellent ratings and recognition based on current environmental performance standards and metrics were rated worse than average, according to the study. But the study failed to take into account complexities such as product mixes, number of sites, and types of emissions. While proponents of weighting and aggregation argue that such efforts can lead to better decision making, critics suggest that rather than more sophisticated metrics, it is more simple and useful metrics that are required. Furthermore, basic management approaches to improving environmental performance have yielded more challenging goals. For example, the goal of continual improvement toward zero emissions is ambitious enough that, in the short term, it has been more effective at improving performance than the development of any weighted or aggregated metrics.

Indeed, efforts to identify, define, and prioritize environmental issues coupled with the development of a publicly accepted understanding of those issues, may be more important than efforts to improve environmental metrics. Such an approach could lead to scientific methods to test for, or at least estimate, the adverse impacts that will guide subsequent industry actions.

## Resource-Use Metrics

Attempts have been made to credit postconsumer waste that substitutes for virgin material, which in the case of the chemical industry is petroleum, or to develop a similar credit for crops and biomass that are used for raw materials. Currently, however, no agreed-upon metrics have been established. Likewise, there is a question as to whether firms should be credited for increasing a product's useful life in instances where the "value" of a chemical is rented but the manufacturer maintains control over its recovery, reprocessing, and reuse. In general, there are unresolved questions about how environmental costs and the benefits of recyclability and retrofitting ought to be measured.

There is also generally no acceptable method of balancing the environmental costs of increased material or energy use in production with the environmental benefits a product may create elsewhere in the economic system. Plastics that reduce the weight of cars and thus increase fuel efficiency are a prominent example.

The practice of normalization by mass is relatively simple, but it fails to measure environmental burden with respect to the function delivered. For example, take a new generation of agrochemicals that accomplishes the same (or better) weed control at about 1/200 the application rate of the previous generation. This means the farmer will handle less material and that there is less material being applied to the land. But, when evaluated on the basis of manufacturing waste or energy used per pound of chemical produced, these new products may appear more material or energy intensive, using current metrics. Their performance on a weight-per-acre-treated basis is superior, but such characteristics are unlikely to be captured in current metrics.

Energy is a vital input in efforts to recover product (and product precursors) through solvent recycling and waste streams. In fact, there is usually an increase in energy use for gains in materials efficiency. While recycling by itself is frequently reported internally and externally, no effective way has been found to capture this trade-off. Metrics that capture the trade-offs between energy and materials use, or between the manufacture of a product and its use, may have the greatest long-term impact. They could be useful not only to customers but also to other industries. Developing such metrics represents a challenging task.

## Sustainability

Several preliminary efforts are under way within the chemical industry to define meaningful metrics relating to sustainable development. These metrics are addressing such issues as energy efficiency, including considerations for energy from truly renewable sources; materials efficiency, including considerations for truly renewable materials; capability for recycling and recycle content; and toxic dispersion corrected for quantified toxicity and for exposure pathway. In each of these areas there is a need to develop a common understanding of the concept of sustainability, develop metrics that drive continuous improvement, garner public acceptance for sustainability metrics, and establish benchmarks against which individual companies can measure themselves.

While there are certainly many issues related to sustainability that will be resolved only through public debate, the chemical industry would benefit by exploring the following issues:

- Use of raw materials and other process inputs derived from renewable and postconsumer recycle sources. Some care must be exercised here, since products made from renewable or recycled materials are not always better from an environmental standpoint. In some cases, the energy demand for the collection system or the product separation steps can be so high as to produce a net $CO_2$ increase comparable with the use of fossil carbon and "virgin" materials.

- Use of renewable energy or other "green" energy.
- Use of reprocessed water, with emphasis on returning spent water in superior condition to aquifers or to potable supplies.
- Feedstocks with lower environmental impact than currently competitive choices.
- Impact of manufacturing processes and product use on such things as biodiversity, habitat loss, and deforestation.

## REFERENCES

American Chemical Society. 1998. Technology Vision 2020: The U.S. Chemical Industry. Available online at http://www.chem.purdue.edu/v2020/. [February 5, 1999]

Carberry, J. Forthcoming. Environmental knowledge systems at DuPont. Paper in Green Tech·Knowledge·y: Information and Knowledge Systems for Improving Environmental Performance. Washington, D.C.: National Academy Press.

Chemical Manufacturers Association (CMA). 1995. CMA Statistical Handbook. Washington, D.C.: CMA.

Chemical Manufacturers Association (CMA). 1998a. Responsible Care: A Public Commitment. Available online at http://www.cmahq.com/cmaprograms/rc/about/content_rc_about.html. [August 10, 1998]

Chemical Manufacturers Association (CMA). 1998b. Responsible Care: Codes of Management Practices: Product Stewardship. Available online at http://www.cmahq.com/cmaprograms/rc/rcnews/lc_rc_rcnews.html. [August 10, 1998]

Imperial Chemical Industries. 1998. Environmental Burden Approach. Available online at http://www.ici.com/download/index.htm. [August 10, 1998]

Kiernan, M., and J. Levinson. 1997. Environment drives performance: The jury is in. Environmental Quality (Winter):1–8.

Magretta, J. 1997. Growth through global sustainability: An interview with Monsanto's CEO, Robert B. Shapiro. Harvard Business Review 75(1):78–83.

Solomon, C. 1993. Clearing the air. What really pollutes? Wall Street Journal, March 29.

White, A., and D. Zinkl. 1997. Green Metrics: A Status Report on Standardized Corporate Environmental Reporting. Boston: Tellus Institute.

# 6

# The Electronics Industry

## BACKGROUND

Electronics, computers, and associated software have transformed every facet of society. In addition to providing the basis for the information revolution, electronics enable many of society's vital support systems, including those that provide for such necessities as food, water, energy, transportation, health care, telecommunications, trade, and finance.

The electronics sector produces a diversity of devices and equipment. Industries in this sector fall under U.S. Department of Commerce Standard Identification Code (SIC) 35 and 36. SIC 35 describes industries that produce electronic computers, computer storage devices, computer terminals, computer peripheral equipment, calculating and accounting equipment, and office machines. SIC 36 describes industries that produce electron tubes, printed circuit boards, semiconductors and related devices, electronic capacitors, electronic coils and transformers, electronic connectors, and electronic components. Due to the great diversity of electronics products and the desire to provide an in-depth analysis, this chapter will focus on those metrics used in the manufacture of semiconductor devices and consumer electronics products.

### The Semiconductor Manufacturing Process

Semiconductor manufacture begins with a solid crystalline material whose electrical conductivity falls between that of metal and insulator. The most common materials used are silicon and germanium. These are processed to produce semiconductors (commonly referred to as integrated circuits, or ICs), defined as miniature electronic circuits produced within and upon a single semiconductor crystal (McGraw-Hill Book Company, 1995). Semiconductors serve two purposes: they act either as a conductor, guiding or moving an electrical current, or as an insulator, preventing the passage of heat or electricity. Typical functions of semiconductors in electronic products include information processing, displays, power handling, data storage, signal conditioning, and converting light energy to electrical energy, or vice versa.

IC manufacturing is complex and involves the use of ultra-high-purity liquids and gases. Semiconductor manufacturing can be broken down into six basic steps (Box 6-1). The primary concern during manufacturing is contamination of the product. All steps are, therefore, carried out in very clean environments that consume as much as 60 percent of the electrical power used in wafer fabrication

---

**BOX 6-1**
**Steps in the Manufacture of Semiconductors**

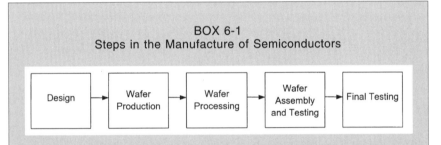

**Step One: Design**

The circuit is designed using computer modeling techniques. A structural description of the design is developed from the given electrical specifications. After the circuit has been designed, the design is verified using computer simulation to test functionality and to develop and test layouts of the circuit path. The layout phase identifies the location of the circuits on the silicon surface and their interconnections. Computer simulation analyzes the completed layout to verify complex geometrical constraints. The designers develop a set of mask descriptions when the layout is complete. A prototype chip is manufactured and returned to the designers for extensive testing, including diagnostic testing in which actual performance is compared with design expectations.

**Step Two: Wafer Production**

A wafer is a thin, round slice of a semiconductor material, usually silicon. In wafer production, purified polycrystalline silicon, created from sand, is heated to a molten liquid. A small piece of solid silicon (seed) is placed on the molten liquid. As the seed is slowly pulled from the melt, the surface tension between the seed and molten silicon causes a small amount of the liquid to rise with the seed and cool. The resulting crystal ingot is then ground to a uniform diameter and a diamond saw blade cuts the ingot into thin wafers. The wafer is processed through a

continued

BOX 6-1 *Continued*

series of machines, where it is ground smooth and chemically polished to a mirror-like luster. The wafers are then ready to be sent to the wafer fabrication area, where they are used as the starting material for manufacturing integrated circuits (Harris Corporation, 1998).

Semiconductors are extremely sensitive to contamination. Airborne particulates, even 1 μm in size, can cause defects. Special precautions are taken in production to reduce the amount of particulates in the air. For example, lint-free garments are worn by the process personnel to minimize operator-induced particulates.

### Step Three: Wafer Processing

Wafers are typically processed in batches of 25 to 40. Semiconductor fabrication is very complex. First, films are deposited or grown, usually through oxidation, on the single-crystal surface, which can serve to provide a protective barrier, can be used as a dielectric, or can serve to isolate devices or layers. The typical next step, usually called photolithography, is one of the most crucial. It is often repeated 8 to 15 times. Photolithography imprints patterns onto the silicon substrate. Incorrect patterning affects the semiconductor's electrical properties. An etching step usually follows to remove selected portions to create patterns. Etching is often followed by a series of steps that introduce controlled amounts of chemical impurity, or dopants, into the film. Dopants are typically used to enhance the semiconducting properties. One of the last steps in the wafer fabrication process is called metallization. One or more layers of a metal alloy are deposited on the wafer surface. This metal provides the physical and electrical contacts with the silicon. A final protective oxide layer is put on the wafer's surface. This layer protects the semiconductor and insulates it from contact with other external metal components.

### Step Four: Wafer Assembly and Testing

The semiconductor is next tested to ensure that it is performing as designed. A drop of ink is placed on semiconductors that do not meet the design specifications. This minimizes packaging costs since nonconforming semiconductors are discarded during assembly operations. This step also provides information on process yields.

Assembly transforms the device into a useable form while protecting its quality and reliability. Semiconductors are assembled by mounting the wafer onto a metal frame, connecting the wafer to metal strips (leads), and enclosing the device to protect against mechanical shock and the external environment. The enclosure can be plastic or ceramic. Today, the majority of devices manufactured are enclosed in plastic.

### Step Six: Final Testing

A final series of tests are performed on the device to evaluate conformance to published specifications.

SOURCES: Adapted from McGraw-Hill Book Company (1995), Texas Engineering Extension Service (1994).

facilities, also known as wafer fabs. Some clean rooms have particle levels as low as one to five per cubic foot of air. By comparison, operating rooms have particle levels of 10,000 to 100,000 per cubic foot, and outside air contains about 500,000 to 1,000,000 particles per cubic foot of air. Hazardous materials such as sulfuric acid, hydrofluoric acid, hydrochloric acid, and phosphoric acid are widely used during processing. Patterns are imprinted and developed on the silicon substrate using organic chemicals.

### Environmental Performance Improvements

The use of metrics for decision making within the semiconductor industry, as for the other industries studied, is generally driven by regulation. One exception is efforts to reduce energy and water use, which generally result from a desire to lower operating costs. The electronics industry is critically dependent on rapid technological innovation, and it is beginning to apply similar efforts to meeting environmental challenges. In the early 1990s the industry began developing a "road map" that identifies research needs for improved semiconductor products and processes. Recently, environmental considerations have been integrated into every aspect of this technological road map.[1] The 1997 road map identifies several environmental challenges. These are summarized in Table 6-1 along with the environmental issues and potential metrics associated with each.

The three most difficult technical challenges identified in the road mapping exercise are to ensure early distribution of information about the toxicity and safety of chemicals to users, reduce water and energy use, and reduce perfluorocarbon (PFC) emissions (Semiconductor Industry Association, 1997). PFCs are used to etch silicon wafers and to clean plasma chambers used in semiconductor manufacture. If released to the atmosphere, these long-lived compounds act as greenhouse gases. Many semiconductor companies have voluntarily entered into a memorandum of understanding with the U.S. Environmental Protection Agency (EPA) to address PFC emissions.

### CURRENT USE OF ENVIRONMENTAL PERFORMANCE METRICS

Metrics help semiconductor companies choose chemicals, processes, or products that have minimal environmental risk. Metrics, such as chemical use, identify processes that are material intensive or that use high-risk chemicals. Energy or natural resource consumption and state regulatory requirements are often considerations in choosing a location for a new facility. Environmental metrics can

---

[1]To ensure that environmental, health, and safety considerations are integrated into the road map, the Semiconductor Industry Association enlists the help of its Safety and Health Committee. This committee monitors, identifies, and addresses priority environmental issues at the federal, state, and local levels.

TABLE 6-1   Environmental Challenges in the Semiconductor Industry

| Challenges | Summary of Issues | Possible Metrics |
|---|---|---|
| *Anticipated Before 2006 (circuitry dimensions ≥ 100 nm[a])* | | |
| New chemical qualification | Need to conduct thorough new chemical reviews and ensure that new chemical processes can be utilized in manufacturing without jeopardizing human health or the environment or delaying process implementation. | Number of new chemical reviews conducted |
| Reduce PFC emissions | These gases are used in plasma processing. There are no known alternatives. International regulatory scrutiny is growing. | PFC emissions |
| Reduce energy and water use | Availability of energy and water may limit location and size of wafer fabrication facilities in certain geographic regions. | Energy and water use, alternative reuse opportunities |
| Integrated ESH impact analysis capability | There is no integrated way to evaluate and quantify the impact of process, chemicals, and process tools. | EHS cost per unit of production |
| *Anticipated After 2006 (circuitry dimensions < 100 nm[a])* | | |
| Eliminate PFC emissions | There are no known alternatives and international regulatory pressure. | PFC emissions |
| Know detailed chemical characteristics before use | Need to document toxicity and safety characteristics because of international regulatory pressure. | Number of risk assessments conducted on new chemicals |
| Lower use of feed water by a factor of 10 and halve cost of water purification | Reducing use and cost of water will improve productivity curve and increase flexibility of factory siting. | Water use, alternative reuse opportunities, cost to provide purer water |
| Halve energy use per unit of silicon | Desire to reduce global-warming impact of energy use. Energy availability in market area. | Energy use |
| Integrated ESH impact analysis capability for new designs. | Lack of an integrated way to make ESH a design parameter in development procedures for new tools and processes | Partnering with manufacturing tool suppliers to develop metrics for cleaner tools |

NOTE:  PFC = perfluorocarbon; ESH = environment, safety, and health.
[a]nm (nanometer) refers to the width of a beam of light from the lithographic light source.  With smaller dimensions of the circuitry (i.e., lines and spaces), narrower beams are required.
SOURCE:  Semiconductor Industry Association (1997).

be a factor in any of these decisions. Success stories involving the use of environmental metrics can work to the benefit of a firm during permit or regulatory negotiations.

Figure 6-1 shows the resource inputs, product outputs, waste, and associated metrics of a standard semiconductor manufacturing facility. Regulated materials, such as Toxic Release Inventory (TRI) chemicals, are often tracked throughout the manufacturing process. Depending upon the characteristics of a company's materials and design capabilities, some materials are recycled through a closed loop within the facility. Others are sent to recyclers or disposed of off site.

## Environmental Burden

Environmental burden in the electronics industry relates primarily to chemicals that have the potential to be released to the air, water, or land. In 1995 the electronics industry accounted for 446.7 million pounds of TRI releases and transfers, of which 15.8 million pounds were associated with the semiconductor industry (Right-to-Know Network, 1998). Since the expected growth rate of the semiconductor market is projected to be 20 percent over each of the next three years (Semiconductor Industry Association, 1997), chemical management is critical to the industry. In addition to TRI emissions, the industry also tracks ozone-depleting substances, EPA 33/50 chemicals, and hazardous waste. Some within the industry also track PFCs.

Chemical management involves monitoring chemicals that are used and released into the environment as well as tracking regulatory inspections and compliance issues. The goal of chemical management is to minimize risks to safety, public and employee health, and the environment. To date, the industry has successfully applied pollution prevention principles to the management of chemicals and emissions (Box 6-2). The industry can also boast of having essentially eliminated Class I ozone-depleting substances from its manufacturing processes.

The industry's ability to use new chemicals depends on a robust chemical assessment and selection process. The challenge lies in developing tools that will assist in the selection of chemicals that meet the needs of semiconductor manufacturing while also improving environmental performance. As part of this effort, it is important for managers and designers to get information about the environmental and health characteristics of potential new process materials as early as possible, thus limiting health risks, environmental liabilities, and potential downtime.

## Resource Use

Beyond the chemical requirements, other primary resource inputs to semiconductor manufacturing are the semiconductor substrate, water, energy, and packaging material. Innovations have allowed an increase in wafer size but at the

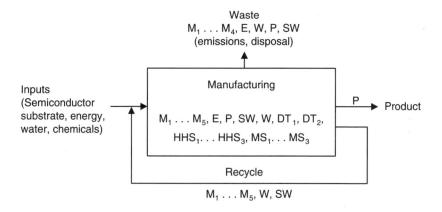

$M_1$ = TRI chemicals
$M_2$ = Ozone-depleting substances
$M_3$ = 33/50 chemicals[a]
$M_4$ = Hazardous waste
$M_5$ = SARA chemicals
E = Energy use (electricity, natural gas, and fuel)
W = Water use
P = Packaging materials
SW = Nonhazardous solid waste
$DT_1$ = Environmental cost accounting
$DT_2$ = Design for environment (e.g., number of environmentally designed products)

$MS_1$ = Environmental management systems
$MS_2$ = Regulatory inspections
$MS_3$ = Compliance issues
$HHS_1$ = Accidents/injuries per 100 employees
$HHS_2$ = OSHA recordable injuries and illnesses
$HHS_3$ = Lost and restricted day cases[b]

FIGURE 6-1  Metrics used in semiconductor manufacturing.
NOTES:  TRI = Toxic Release Inventory; SARA = Superfund Amendments and Reauthorization Act; OSHA = Occupational Safety and Health Administration.
[a]The U.S. Environmental Protection Agency's (EPA's) 33/50 program (also known as the Industrial Toxics Project) is a voluntary pollution reduction initiative that targets releases and off-site transfers of 17 high-priority toxic chemicals.  Its name is derived from its overall national goals—an interim goal of 33 percent reduction by 1992 and an ultimate goal of a 50 percent reduction by 1995, with 1988 being established as the baseline year. The 17 chemicals are from EPA's Toxic Release Inventory.  They were selected because they are produced in large quantities and subsequently released to the environment in large quantities and are generally considered to be very toxic or hazardous, and the technology exists to reduce releases of these chemicals through pollution prevention or other means.  Although the goals have been met—a 40 percent reduction was achieved by 1992, and 50 percent reduction was reached ahead of schedule in 1994 (United States Environmental Protection Agency, 1999)—companies continue to track these 17 chemicals.
[b]Lost and restricted days are those in which a worker is unable to perform a particular function due to illness or injury but is able to perform other tasks.

BOX 6-2
Pollution Prevention Efforts Reduce Emissions

Organic solvents are frequently used to remove contaminants from wafer surfaces in semiconductor manufacturing. Typical cleaning processes involve the use of industry standard equipment that specifies parts placement and dictates how solvents are to be applied.

Through innovative parts placements and modification of the solvent application technique, engineers at IBM Burlington (Vermont) were able to substantially increase the number of parts processed per batch and improve cleaning efficiency. The new process reduced the site's solvent use by 1,860 metric tons in 1996 and saved over $5 million in chemical and production costs.

Similarly, through the development and implementation of no-clean fluxes in three processes, IBM's Bromont (Canada) facility has achieved a 70 percent reduction in its perchloroethylene (perc) emissions since 1993. The plant's goal is to completely eliminate perc emissions by the end of 1998.

SOURCE: International Business Machines (1998).

cost of more process steps. This has resulted in a need for water of greater purity and more water use per wafer. Increased water needs can be met by a combination of strategies, including higher-efficiency rinse processes, recycling of higher-quality water for process applications, and reuse of lower-quality water for nonprocess applications. At Intel, 50 to 70 percent of industrial water is recycled ultrapure water (Intel, 1998). Typical metrics for water use are gallons per year or gallons per employee per day. Wastewater reuse may be tracked as total gallons or as a percent of total wastewater.

The industry also faces a challenge in the area of energy use. In 1995 the U.S. semiconductor companies consumed a total of 8.4 billion kWh of electricity (Semiconductor Industry Association, 1997). Table 6-2 shows operating expenses of a typical semiconductor facility. The electric bill can be the largest or second-largest expense item, representing 25 to 40 percent of a facility's operating budget (excluding capital and construction expenditures; Semiconductor Industry Association, 1997).

## Mounting Components and Packaging

Once a semiconductor assumes final form, it is usually mounted onto a circuit board. Circuit boards often contain many semiconductors and capacitors. Epoxy compounds are frequently used to attach components to the board. Although excellent at reinforcing the interface between components and substrates,

TABLE 6-2   Operating Expenses of a Typical Semiconductor Facility

| Category | $1,000s | Percent of Total Expenses |
|---|---|---|
| Central plant | 3,720 | 7.1 |
| Building/structure | 3,838 | 7.3 |
| Ultrapure water | 539 | 1.0 |
| Chemical services | 479 | 0.9 |
| Gas services | 247 | 0.5 |
| Electricity | 21,890 | 41.5 |
| Water | 1,423 | 2.7 |
| Natural gas | 2,481 | 4.7 |
| Custodial (cleaning service) | 1,953 | 3.7 |
| Landscaping | 102 | 0.2 |
| Trash | 158 | 0.3 |
| Hazardous waste | 774 | 1.5 |
| Salary/benefits | 15,105 | 28.7 |

SOURCE: Allenby (forthcoming).

these polymers make it difficult to recover portions of assemblies or remove and recycle chips and components. Some in the electronics industry, such as consumer and office equipment manufacturers, are beginning to recycle their products, a trend that is driving IC makers to find new mounting materials. Companies that produce both semiconductors and consumer electronics products have addressed this by developing alternative adhesive compounds. IBM Research, for example, has developed a new epoxy compound that is easily removed, allowing for rework or disassembly of components for recycling. The new epoxy can be dissolved in specially designed, water-based, mildly acidic systems, yet it still meets all the performance requirements of a typical epoxy (Buchwalter and Kosbar, 1996). This is an example of how design for environment[2] (DFE)

---

[2]DFE is an approach to implementing environmental design programs within the concurrent engineering framework that many electronics companies use in their product realization process. As defined by Winner et al. (1988), concurrent engineering

is a systematic approach to the integrated concurrent design of products and their related processes, including manufacturing and support. This approach is intended to cause the developers, from the outset, to consider all elements of the product life cycle from conception through disposal, including quality, cost, schedule, and user requirements.

Thus, integrating environmental considerations in concurrent engineering—the essence of DFE efforts—ensures that environmental factors are taken into account during development and design stages and not left to be dealt with after vital decisions are made.

practices can positively affect the recyclability and longevity of a product, thereby saving money.

Increasing production volumes are driving companies to look for ways to reduce the amount of materials being used, including shipping containers. Many companies have identified opportunities to minimize or to reuse packaging material. Texas Instruments (TI) has successfully reduced packaging by 70 percent and increased shipping capacity by 33 percent. TI also worked with one of its customers, Ford Motor Company, to develop packaging material that is reusable or made from recyclable material. Today, 91 percent of packaging material shipped between TI and Ford is recyclable or reusable (Texas Instruments, 1998).

## Human Health and Safety

As is true for many other industries, worker health and safety is tracked in terms of accidents and injuries per 100 employees, OSHA recordable injuries and illnesses, and lost or restricted day cases.[3] The Semiconductor Industry Association road map suggests that in the future more attention will be focused on improving manufacturing equipment for semiconductors and the selection of process chemicals to provide additional worker protection. Part of the chemical selection process will involve the use of risk assessment and risk management procedures. There will, however, still be an ongoing need to identify reliable, cost-effective tools to monitor work areas for potential exposure to toxic chemicals. Other issues related to worker protection that must be considered include improved information flow at worksites to ensure that accurate, appropriate information is disseminated to all employees; increased attention to maintenance operations; and better understanding of the risks and implications of physical hazards in the workplace.

## Design Approach

Semiconductor manufacturers tend to focus on cost, yield, and logistics when selecting products and processes. However, because of the costs of coping with them, environment, safety, and health (ESH) risks are also becoming important drivers in design and operational decision making. When not considered initially, ESH issues have resulted in major postinstallation changes to processes and increased operating costs. To reduce ESH-related costs, risk factors must be evaluated and dealt with at an early stage in the design process. The DFE approach is intended to produce products that are environmentally acceptable throughout their life cycle.

---

[3]Lost or restricted days are those in which a worker is unable to perform a particular function due to illness or injury but is able to perform other tasks.

In semiconductor manufacturing, DFE usually entails

- designing out the source of or the regulatory requirement for the environmental concern,
- using design options that pose less environmental risk,
- using fewer materials and processes to reduce waste,
- minimizing materials or processes that may create an environmental risk, and
- using design constraints and engineering controls that reduce the potential for undesirable environmental effects or releases.

These principles ensure that the environmental consequences of a product's life cycle are understood and addressed before or as manufacturing decisions are made. Design teams use checklists, guidelines, Web pages, and supplier specifications to choose processes and product features that are less harmful to the environment. The intent is to produce the most environmentally sound product design that will also meet function, cycle time, cost, and quality goals. Progress toward meeting these goals is tracked through specific metrics associated with each principle.

## Summary

Metrics are often used as indicators to guide the design or retrofitting of a facility or manufacturing operation. They can also be used to forecast operational and risk-based expenses, track facility performance, establish goals and targets for a corporation, and create a baseline of company performance that can be used to benchmark against other companies in the industry.

To examine the use of environmental performance metrics by semiconductor companies, the committee surveyed member companies of SEMATECH (SEmiconductor MAnufacturing TECHnology), a nonprofit research and development consortium of U.S. semiconductor manufacturers.[4] using publicly available information, such as websites and environmental annual reports. A summary of the results is shown in Appendix B. The survey illustrates that even within an industry sector the types of environmental metrics that are reported vary depending on the structure of the company and its products. IBM and Digital, for example, produce more than semiconductors and therefore report on their performance in terms of a full range of products. Consequently, their environmental reports are not readily comparable with a company like Intel,

---

[4]Companies surveyed were Advanced Micro Devices, Digital Equipment, Hewlett-Packard, Intel, IBM, Lucent Technologies, Motorola, National Semiconductor, Rockwell International, and Texas Instruments.

whose primary product is integrated circuits. Meaningful comparisons can, therefore, be difficult even within a single industry.

A compilation of environmental metrics used by semiconductor manufacturers appears in Table 6-3.

## CHALLENGES AND OPPORTUNITIES: SEMICONDUCTORS

Semiconductor technology changes very rapidly. ICs are decreasing in size as their performance and capacity increase. At the same time, wafer size is increasing, allowing more of the smaller ICs on each wafer. Moore's Law, which predicts that IC performance will double every 18 months, is expected to hold true through 2010 (Semiconductor Industry Association, 1997). Because of this rapid pace of change and the industry's concurrent engineering practices, IC manufacturers tend to be adapt quickly. As a result, the industry is able to adjust to new environmental factors more rapidly than industries that have products and processes with longer life cycles. Applying DFE to effect such changes, however, is relatively new, and the tools and techniques being used appear to be somewhat rudimentary, often involving the use of guidelines and checklists. Efforts to build more sophisticated tools and have them adopted by companies have not always met with great success (Hoffman and Scheller, 1998).

Currently, most semiconductor companies are not factoring the recyclability of the final products into design processes. Efforts like IBM's development and use of epoxy alternatives, however, give some cause for optimism.

The lack of comparable metrics is an important issue in the industry. Unlike financial reports, where metrics are standardized, there is no uniformity in the type of information included in company environmental reports or in the units used to report it. Standardization is needed for meaningful comparisons, but until the financial community demands such standardization, it is unlikely to happen. Meanwhile, there are reports in the financial media that some investment fund managers are using environmental performance to augment the traditional screening process used to rank companies (Deutsch, 1998). If this practice becomes commonplace, it will likely bring greater standardization to industrial environmental performance metrics. Efforts by grassroots organizations to get companies to comply with certain standards, such as principles developed by the Coalition for Environmentally Responsible Economies (CERES), through stockholder proposals are becoming more common.[5]

The International Organization for Standardization (ISO), particularly through ISO 14031, has sought to provide some much-needed definitional and

---

[5]At a recent Intel annual meeting, the company considered a proposal by stockholders to have the company abide by the CERES principles. Although the effort failed, it shows a trend of stockholder activism.

TABLE 6-3    Environmental Metrics Used in Semiconductor Manufacturing

| Manufacturing | Product Use |
|---|---|
| *Resource Related* | |
| Chemical Management | Natural Resources |
| • TRI emissions | • Packaging materials |
| • ODS chemicals | |
| • 33/50 chemicals[a] | Design Tools |
| • Hazardous waste | • DFE |
| • Global-warming chemicals | • Environmental cost accounting |
| • Natural resources | |
| • Energy use | |
| • Water use | |
| • Packaging materials | |
| | |
| Design Tools | |
| • DFE | |
| • Environmental cost accounting | |
| | |
| *Environmental Burden Related* | |
| Chemical Management | Natural Resources |
| • Regulatory issues | • Packaging |
| e.g., inspections, audits | • Landfill disposal |
| • Compliance issues | |
| e.g., fines, violations | |
| • Hazardous waste | |
| • Superfund | |
| • Remediation | |
| | |
| *Human Health and Safety* | |
| Worker Protection | |
| • Accidents/injuries per 100 employees | |
| • OSHA recordable injuries and illnesses | |
| • Lost and restricted day cases[b] | |

NOTE: TRI = Toxic Release Inventory; ODS = ozone-depleting substances; DFE = design for environment.

[a]The U.S. Environmental Protection Agency's (EPA's) 33/50 program (also known as the Industrial Toxics Project) is a voluntary pollution reduction initiative that targets releases and off-site transfers of 17 high-priority toxic chemicals. Its name is derived from its overall national goals—an interim goal of 33 percent reduction by 1992, and an ultimate goal of a 50 percent reduction by 1995, with 1988 being established as the baseline year. The 17 chemicals are from EPA's Toxic Release Inventory. They were selected because they are produced in large quantities and subsequently released to the environment in large quantities; they are generally considered to be very toxic or hazardous; and the technology exists to reduce releases of these chemicals through pollution prevention or other means. Although the goals have been met—a 40 percent reduction was achieved by 1992, and 50 percent reduction was reached ahead of schedule in 1994 (United States Environmental Protection Agency, 1999)—companies continue to track these 17 chemicals.

[b]Lost and restricted days are those in which a worker is unable to perform a particular function due to illness or injury but is able to perform other tasks.

comparability guidelines to industry. The basic premise of ISO 14031 is that business strategy should drive metrics rather than the other way around. Environmental metrics should, therefore, reflect the nature and scale of the operations, and selected metrics should provide managers with sufficient information to evaluate progress toward environmental goals. In addition, the ISO standards suggest that in some instances quantitative metrics may be substituted by qualitative indicators (such as those discussed in Chapter 10).

## THE ELECTRONICS LIFE CYCLE

Semiconductors are critical to the operation of virtually all electronics, although they account for only a small portion of sales for the nearly $400 billion U.S. electronics industry. The semiconductor industry and related high-technology industries now account for 30 percent of America's economic growth. Fifty percent of semiconductors are used in computers, 17 percent are used in consumer electronics, and the remainder are installed in cars, communications systems, industrial applications, instruments, and defense systems (Semiconductor Industry Association, 1998). A simplified heirarchy of the electronics industry is shown in Figure 6-2.

The life-cycle concerns of each segment of the industry vary widely. For example, in semiconductor manufacturing, environmental concerns are primarily at the process level. They include PFC use and recovery, hazardous waste use and management, chemical disposal, water and energy use, and in-process recycling. Product take-back and stewardship are not much of a concern, since semiconductors are integrated into larger products, which themselves may be the subject of corporate environmental stewardship efforts.

Printed circuit board manufacturers, like semiconductor manufacturers, are concerned with the use and disposal of hazardous materials. Their products are typically the components of consumer electronic products and may be returned to them at some future date. Hence, these firms might also be concerned with heavy metals and recycling of scrap boards and metals.

Manufacturers of consumer electronics products, on the other hand, are like automobile manufacturers: They have to take into account consumer as well as

```
                                                              Trains
                                                              Planes
                                                              Automobiles
Semiconductor ⇨  Printed Circuit Board ⇨  Electronics ⇨      Computers
                                                              Appliances
                                                              Toys
                                                              Etc.
```

FIGURE 6-2   Heirarchy of the electronics industry.

environmental factors.  In general, their environmental objectives are related to energy efficiency, minimizing electromagnetic signal emissions, and ensuring that there are no measurable outgasing or toxicity hazards from materials and components of the electronic product.  Materials used in the product must meet safety standards and emit no toxics in case of fire.  Plastics are increasingly being used because of their light weight and low cost, but the use of halogenated plastics and fire retardants is a concern because of their potential to form dioxin during a fire.  Use of  substances such as cadmium in batteries also poses risk to the environment.  This segment of the industry is beginning to establish recycling centers to take back their products for reuse or recycling (Box 6-3).

### Life-Cycle Studies and DFE Metrics

Studies of electronics products have revealed that environmental impacts occur throughout their life cycles.  Given the rapid change in the industry, the accuracy of these data is short lived.  For example, the results of a much-quoted study (Box 6-4) were from data collected in the early 1990s.  The computer workstation that served as the focus of the study has changed dramatically, making specific results less relevant, but trends related to product energy use are still valid.

A growing body of literature on DFE in the electronics industry suggests several common practices related to life-cycle factors and waste-stream issues in the disposal of products.  Broader concerns about the end of product life are evident in DFE principles, which include

---

**BOX 6-3**
**Computer Equipment Recycling Centers**

Digital Equipment Corporation, which recently merged with Compaq, operates two computer recovery centers—one in Contoocook, New Hampshire, and the other in Nijmegen, Netherlands.

At the centers, incoming computer equipment is first inspected to determine the best approach for recovery. Usable equipment is repaired, refurbished, or sold for reuse. Unusable and obsolete equipment goes through a disassembly operation. Generic components with commercial value such as integrated circuits, memory chips, and disk drives are extracted and sold for reuse. The remaining assemblies are then dismantled and separated. These groups of materials are dispatched to specialized vendors for recycling or disposal using controlled, prequalified processes.

SOURCE: Digital Equipment Corporation (1998).

## BOX 6-4
## Environmental Profile of a Computer Workstation

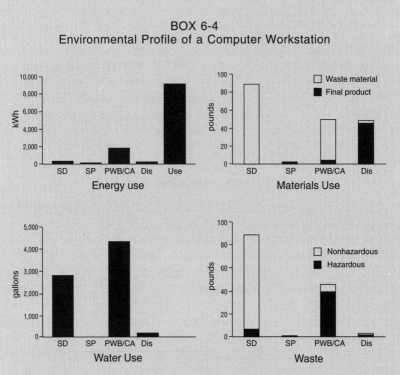

Results of a life-cycle study of the computer workstation are summarized in the graphs above. The computer workstation studied was assumed to contain one $\frac{1}{6}$-inch-thick silicon wafer (about 28 square inches), 220 integrated circuits (213 in plastic and 7 in ceramic packages), about 500 square inches (3.6 square feet) of single and multilayer printed wiring board, and a 20-inch monitor. The subcomponents included in the study were semiconductor devices (SD), semiconductor packaging (SP), printed wiring boards and computer assemblies (PWB/CA), and display units (Dis). The profiles of energy, material, and water use and waste reveal some aspects of the environmental impacts of an electronics product.

### Energy Use

As in the case of the automobile, the greatest energy is consumed on a per-product basis during computer use. Although energy consumption during use continues to dominate the energy profile of the computer, several more recent models feature components that require less power and that "power down" when inactive. EPA recognizes such products through its Energy Star program. which many manufacturers use for product marketing purposes.

### Materials Use

Both product and process materials are used to manufacture a computer. Product materials become part of the product, while process materials are used in

continued

the processing of parts and components but do not end up in the product. Process materials include various gases and cleaning solvents. The processing of various computer subcomponents generates differing quantities of waste. The production of semiconductor devices and printed wiring boards is the most materials-intensive portion of the computer manufacturing process. The production of printed wiring boards involves several lithographic, plating, and etching processes that require significant amounts of chemicals. The display terminal uses the least amount of process materials but is the greatest contributor to the weight of the computer.

**Water Use**
Water is critical to the manufacture of the various computer subcomponents. Substantially more water is used in the manufacture of printed wiring boards than semiconductor devices and other subcomponents. Water used per unit of product is greater for semiconductor manufacture than for printed circuit board manufacture.

**Hazardous and Nonhazardous Waste**
Hazardous and nonhazardous wastes are residuals resulting from the manufacture of computers. Printed circuit board production results in the greatest amount of hazardous waste among the four manufacturing processes evaluated in this study.

SOURCE: Microelectronics and Computer Technology Corporation (1993).

- design for disassembly and separability, or simplifying product disassembly and material recovery using techniques such as color-coding plastics or snap fasteners to hold components together;
- design for recyclability, or ensuring both high recycled content in product materials and maximum recycling so there is minimum waste at the end of product life;
- design for reusability, or ensuring that components are compatible with different product lines and recovered, refurbished, and reused across product lines;
- design for remanufacture, or enabling recovery of postindustrial or postconsumer materials for recycling as input to the manufacture of new products; and
- design for disposability, or ensuring that all materials and components can be safely and efficiently disposed of.

Electronics companies with large, leased-based products have also begun to incorporate another practice: product life extension. The practice reduces product or component materials that end up in the waste stream. The idea is to provide and market the functionality of the product in lieu of new hardware (Stahel, 1997). In addition to the DFE practices outlined above, this approach requires making sure that parts common to several products are designed to be interchangeable, and it necessitates managing the logistics and distribution of leased products in the marketplace. One electronics company that is aggressively applying this practice is Xerox (Box 6-5). An appropriate metric in this circumstance would be "life of a part or product." There are limits to this practice, however, as older stock cannot always be upgraded to the newer digital technologies. What is emerging in the electronics industry is a flexible system of product management. A variety of metrics have merged along with these practices.

Common metrics used in the electronics industry are summarized in Box 6-6. They are not dissimilar to those used in semiconductor manufacturing, except that they include metrics related to product use and disposal.

## CHALLENGES AND OPPORTUNITIES: CONSUMER ELECTRONICS

DFE-related metrics and those driven by regulations are useful in the design phase of a product. Individually, however, they are not representative of the environmental performance of a company, except in specific instances such as energy use or release of TRI chemicals. Global environmental metrics are not readily available today. Indeed, the many ways in which the uses of electronics lead to environmental improvement and contribute to sustainable development are not captured in the current set of environmental metrics. Meeting the needs of the present without compromising the needs of the future is the thrust of sustainability, and industrial environmental performance is an important barometer of success. Cleaner air, cleaner water, and reduced exposure to toxics all indicate progress toward sustainable development.

Electronics, a vital part of the telecommunications and computer revolutions, have and continue to transform industrial production and management throughout the economy (Freeman, 1992). The impacts of this revolution on improving environmental performance are already being felt, particularly in the monitoring and control of energy emissions and materials use, in aiding quality and inventory controls, and through improved control of manufacturing processes. Many energy-saving technologies and process changes that promote cleaner production depend on the incorporation of electronic sensors and monitors. System models of production processes are often complicated and their use requires computers. Sensing and monitoring instruments provide essential inputs to the models making it possible to achieve many regulatory objectives. The list of technological advances on the horizon is endless. Electronics-based applications that are likely to reduce societal energy demands include smart buildings, telecommuting, lightweight electronic materials (e.g., electric "paper"), intelligent transportation systems, and improved air traffic management (World Resources Institute, 1998).

## BOX 6-5
## Asset Management and DFE at Xerox

Xerox has adopted modular design as a way of reducing raw material use and waste. Modular design allows for similar parts to be interchanged among several product lines. This makes it easier for the company to recover its leased products and refurbish them for re-lease or resale. By marketing and selling the functions of its products (through leases), the company is in essence managing its leased products as inventory. Part of that management involves recovering usable and interchangeable parts, as shown below.

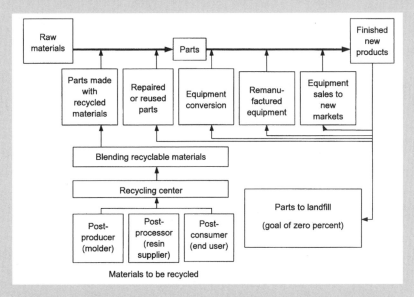

Materials to be recycled

To optimize its recovery of material assets, Xerox has changed its product delivery process; applied DFE practices of disassembly, material recycle, life extension, commonality of parts, and remanufacture and conversion; and developed effective processes for the recovery of used products.

As a result of these efforts, the company has reduced solid waste generation by 73 percent, increased the factory recycle rate by 141 percent, reduced releases to the environment by 94 percent, and realized over $200 million in annual savings.

SOURCE: Calkin (1998).

BOX 6-6
Environmental Performance Metrics Used in the Consumer
Electronics Industry

**Environmental Burden**
- Toxic or hazardous materials used in production
- Total industrial waste generated during production
- Hazardous waste generated during production or use
- Air emissions and water effluents during production
- Greenhouse gas emissions

**Resource-Use Metrics**
- Total energy consumed during product life
- Renewable energy consumed during product life
- Power used during operation (for electrical products)
- Percentage of recycled materials used as input to product
- Percentage of recyclable materials available at end of product life
- Percentage of product recovered and reused
- Purity of recovered recyclable materials
- Percentage of product disposed or type of disposal
- Percentage of packaging or containers recycled
- Useful operating life
- Product disassembly and recovery time

**Economics**
- Average life-cycle cost incurred by the manufacturer
- Purchase and operating cost incurred by the customer
- Cost savings associated with design improvements
- Percentage of products that are leased

Capturing such complex issues in sustainability metrics is a formidable challenge for the future.

## REFERENCES

Allenby, B.R. Forthcoming. The information revolution and sustainability: Mutually reinforcing dimensions of the human future. In Green Tech•Knowledge•y, D.J. Richards, ed. Washington, D.C.: National Academy Press.

Buchwalter, S.L., and L.L. Kosbar. 1996. Cleavable epoxy resins: Design for Disassembly of a thermoset. Journal of Polymer Science and Polymer Chemistry 34(2):249–260.

Calkin, P. 1998. Encouraging modular design. Paper presented at National Research Council/National Academy of Engineering Workshop on Materials Flows Accounting of National Resources, Products, and Residues in the United States, January 26–27, Washington, D.C.

Digital Equipment Corporation. 1998. Environment, Health, and Safety Features of DIGITAL Commercial Desktop Personal Computers. Available online at http://www.windows.digital.com/ resources/whitepapers/ehs%5Fdesktop.asp. [August 10, 1998].

Deutsch, C. 1998. For Wall Street, increasing evidence that green begets green. New York Times. July 19. Section 3, p. 7.

Freeman. 1992. Economics of Hope—Essays on Technical Change, Economic Growth and the Environment. London: Pinter Publications.

Harris Corporation. 1998. How Semiconductors Are Made. Available online at http:// rel.semi.harris.com/doc/lexicon/manufacture.html. [July 30, 1998].

Hoffman, W., and H. Scheller. 1998. Design for Environment at Motorola. Paper presented at NAE Workshop on Industrial Environmental Metrics, January 28–29, Washington, D.C.

Intel. 1998. Water Conservation. Available online at http://www.intel.com/intel/other/ehs/ may97report/water.html. [August 10, 1998].

International Business Machine. 1998. Environment: Pollution Prevention. Available online at http: //www.ibm.com/ibm/environment/annual97/prevent.html. [August 10, 1998].

McGraw-Hill Book Company. 1995. McGraw-Hill Encyclopedia of Science and Technology. New York: McGraw-Hill Book Company.

Microelectronics and Computer Technology Corporation (MCC). 1993. Environmental Consciousness: A Strategic Competitiveness Issue for the Electronics and Computer Industry. Austin, Tex.: MCC.

Right-to-Know Network. 1998. Toxic Release Inventory Database, 1995. Available online at http:// www.rtk.net/www/data/tri_gen.html. [August 10, 1998].

Semiconductor Industry Association (SIA). 1997. The National Technology Roadmap for Semiconductors. San Jose, Calif.: SIA.

Semiconductor Industry Association (SIA). 1998. Annual Report. San Jose, Calif.: SIA.

Stahel, W. 1997. The functional economy: Cultural and organizational change. Pp. 101–116 in The Industrial Green Game, D.J. Richards, ed. Washington, D.C.: National Academy Press.

Texas Engineering Extension Service. 1994. Semiconductor Processing Overview. Bryan, Tex.: Texas A&M University System.

Texas Instruments. 1998. Taking Responsibility for Our Products. Available online at http:// www.ti.com/corp/docs/esh/responsib.htm. [August 10, 1998].

United States Environmental Protection Agency (USEPA). 1999. 33/50 Program: The Final Record. EPA 745-4-99-004. Office of Pollution Prevention and Toxics. Washington, D.C.: USEPA. Also available online at http://www.epa.gov/opptintr/3350/33fin00.htm. [May 19, 1999].

Winner, R.I., J.P. Pennell, H.E. Bertrand, and M.M. Slusarczuk. 1988. Role of Concurrent Engineering in Weapons Systems Acquisition. Report IDA-R-338. Alexandria, Va.: Institute for Defense Analysis.

World Resources Institute. 1998. Taking a Byte Out of Carbon: Electronics Innovation for Climate Protection. Available online at http://www.wri.org/cpi.carbon. [August 10, 1998].

# 7

# The Pulp and Paper Industry

## BACKGROUND

The U.S. paper industry leads the world with over 24 percent of global paper production capacity (American Forest and Paper Association, 1998a). It produces 9 million tons of pulp each year (United States Environmental Protection Agency, 1997a). The pulp and paper industry is the most capital intensive in the United States, spending approximately $130,000 per employee each year in plant and equipment. Economies of scale thus are critical to profitability. Pulp and paper mills produce up to 5,000 tons of paper per day to satisfy a national consumption rate of 700 pounds per American per year, double the consumption in 1960[1] (Blum et al., 1997).

Paper and paperboard products are made from pulp. Pulp is made predominately from wood, but in many cases it is made from other plant fibers such as cotton, linen, and hemp and grasses such as straw, wheat, and kenaf. More recently, recycled paper has become a common material input. In 1996, American paper mills recovered 44.8 percent of postconsumer U.S. paper, and the industry has set a 50 percent paper recovery goal by the year 2000 (American Forest and Paper Association, 1998b).

The industry is relatively well integrated, with some companies managing almost every aspect of the paper cycle: fiber production in forests, pulping, paper

---

[1]Global consumption of paper is currently around 270 million metric tons, three times what it was in 1960 and in line with growth of the global gross domestic product (Grieg-Gran et al., 1997).

making, and paper recycling. In assessing the pulp and paper sector, the committee elected to examine the entire life cycle of the industry (Figure 7-1).

## Forestry

About 3.5 billion cubic meters of wood is harvested worldwide each year (Food and Agriculture Organization, 1995), of which 500 million cubic meters (or 14 percent) is used for pulp and paper, 31 percent for fuelwood, and the rest for solid wood. Data for 1993 indicate that about 650 million cubic meters of wood entered the U.S. economy. Seventy-eight percent was from trees, 10 percent from recycling, and 12 percent from imports. As Figure 7-2 shows, 26 percent was consumed as solid wood, 26 percent as paper, and 36 percent as fuel; 10 percent was exported.

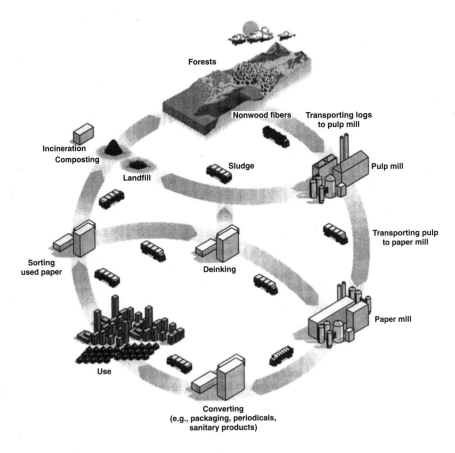

FIGURE 7-1 The paper cycle. SOURCE: Grieg-Gran et al. (1997).

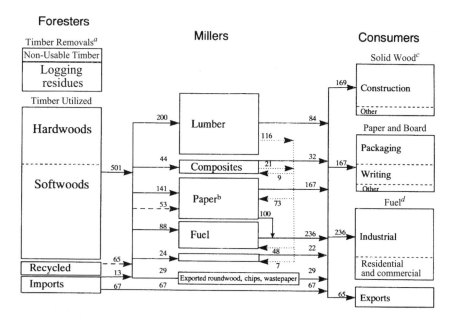

FIGURE 7-2 Material flows in the U.S. forest products industry, 1993. Box heights are to scale. All values in millions of cubic meters (m³). For paper 1 metric ton is equivalent to 2 m³. SOURCE: Wernick et al. (1997).

[a]Based on the ratio of logging residues (15.1 percent) and "Other Removals" (6.6 percent) to all removals for 1991.

[b]The dashed line entering paper from left represents the inputs from "recycled." An estimated 100 million m³ of the woody mass entering paper mills is burned for energy. In 1991 the paper industry (SIC 26) generated over 1.2 quadrillion Btus from pulping liquors, chips, and bark.

[c]Construction includes millwork such as cabinetry and moldings. "Other" includes industrial uses such as materials handling, furniture, and transport.

[d]The ratio of end uses relies on Btu data from the United States Department of Energy, Energy Information Administration. The category "Residential and Commerical" includes electric utilities.

While natural forests are the primary source of pulp wood and solid wood, the global trend is toward acquiring wood grown on plantations or intensively managed natural regeneration forests that resemble plantations. Survey findings on the sources of wood fiber for the global paper industry reveal that 66 percent was from managed natural regeneration forests and plantations, 17 percent from unmanaged regeneration forests, and the rest from virgin sources (Grieg-Gran et al., 1997).

## Pulp and Paper Production

Figure 7-3 shows the production of paper from wood pulp. Logs are first debarked. Stripped bark is then used for fuel or to enrich soil. Wood can be broken down into fibers by mechanical or chemical methods. In the mechanical process, wood fiber is physically separated from the wood by forcing debarked logs and hot water between enormous rotating steel discs with teeth that tear the wood apart or by pressing the logs against grindstones. The mechanical pulping process uses considerable amounts of electricity (2,000 kWh/ton of pulp) but has a high yield (about 90 percent). The wood requirements for mechanical pulping are less than in chemical pulping. Most of the electrical energy that goes into the refiner is liberated as steam, which is subsequently used to dry the paper. Mechanical pulp mills use about 8,000 gallons of water per ton of pulp produced. Bleaching of mechanical pulps is done with hydrogen peroxide or sodium hydrosulfite. Mechanical pulp constitutes about 10 percent of the pulp made in the United States. Recent technology has permitted the construction of mechanical pulp mills that have no liquid effluent. Their only waste is solid waste such as boiler ash. Wood that is chemically processed is chipped. The chips are passed through vibrating screens. Oversized chips and undersized particles (such as chips and dust) are discarded. Accepted chips are stored in huge bins ready for chemical processing. Most pulp produced in the United States is made with the Kraft chemical pulping process (Box 7-1).

To produce paper, the pulp is then screened, cleaned, and mechanically refined. Bales of pulp are dispersed in a huge volume of water so that the slurry is less than 1 percent fiber. The slurry is pumped through a narrow aperture onto a moving wire. The water drains (and is pulled) through the wire to produce a wet pulp mat. Water pulled through the wire is recycled. The wet pulp mat is pressed to remove more water and then is dried over a series of hot rolls to become paper. The water requirements of a paper machine are modest, about 5,000 gal/ton of paper, as much of the water is recycled. The energy requirement, mainly for drying the paper, is roughly 5,000 MJ/ton of paper. This energy often comes from other parts of the mill, such as the mechanical refiners in a mechanical pulp mill or the recovery or waste wood (hog fuel) boilers in a Kraft mill.

## Paper Production Recycling

Paper recycling is on the rise, as shown in Figure 7-4. To recycle fiber, the paper is slurried in water and then run through cleaning and screening operations to remove such contaminants as wire, plastics, paper clips, and staples. In some mills the pulp is also deinked. In mills without deinking processes, excess paper-machine white water is sufficient to run the recycle mill. With deinking, fresh water makeup is required. The volume of fresh water makeup can vary widely, from 1,500 gal/ton of pulp to almost 20,000 gal/ton of pulp, depending on the deinking system (Simons, 1994).

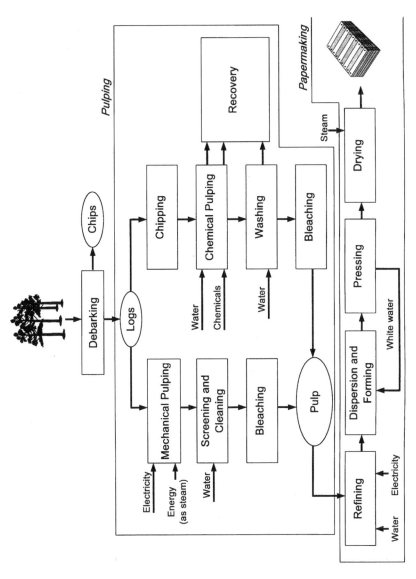

FIGURE 7-3  Pulp production and paper making.

BOX 7-1
Kraft Pulping Process

In the Kraft pulping process, wood chips are fed into a digestor where aqueous sodium hydroxide and sodium sulfide, under heat and pressure, break down the lignin that holds wood fibers together. After pulping, the wood becomes individual fibers. Kraft pulping is an energy-intensive process requiring about 9,000 MJ to produce a ton of pulp. The pulp yield of the Kraft process is about 50 percent. This means that about twice as much wood is required to produce a ton of chemical pulp as is needed to produce a ton of mechanical pulp. The chemical and energy recovery of the Kraft process is efficient, however. Over 98 percent of the pulping chemicals are regenerated in the recovery process. The organic matter that is dissolved during pulping (the other 50 percent of the wood that doesn't become pulp) is fired in a recovery furnace. The energy liberated by burning the dissolved organics—typically 14,000 MJ/ton of pulp—is sufficient to run the pulp mill, with some left over. Many mills sell the excess electricity.

Kraft pulp may be used in an unbleached or bleached form. Bleaching is usually done with oxygen, chlorine dioxide, and peroxide. Total chlorine free (TCF) sequences, which replace the chlorine dioxide with peroxide and often ozone, are used but are not as common as the elemental chlorine free (ECF) sequences. In the United States only one pulp mill is TCF. The recent promulgation of the "cluster rule"[1] will guarantee that ECF sequences become standard in the decade to come. Current technology does not permit bleach effluent to be recycled. If it is sent through the recovery process, salts build up leading to corrosion and scaling. Hence, the bleach plant accounts for fully half of the effluent that comes from the Kraft mill. Total water use in a Kraft mill is about 20,000 gallons/ton of pulp.

---

[1]The cluster rule is a multimedia regulation that requires pulp and paper mills to meet baseline limits for toxic releases to air and water. The limits are expected to eliminate dioxin discharges and cut toxic air emissions by almost 160,000 tons annually. The pulp and paper mill cluster rule is the first issued by the Environmental Protection Agency to control the release of pollutants to two media—air and water—from a single industry. The rule allows pulp and paper mills to select the best combination of pollution prevention and regulatory requirements at one time. It also provides incentives for mills to adopt advanced pollution control technologies that lead to further reductions in toxic discharges. Mills volunteering for this program will be subject to more stringent reductions but will receive rewards for their participation, such as additional compliance time (United States Environmental Protection Agency, 1997b).

## Drivers of Environmental Performance Improvements

As with the other manufacturing industries examined by the committee, regulation has been the dominant driver of environmental performance improvements in the pulp and paper sector. Until the 1980s, the industry's environmental focus was primarily on manufacturing. In the 1980s and 1990s, however, the industry came under additional pressures to improve its environmental performance as concerns related to unsustainable natural resource use, industrial pollu-

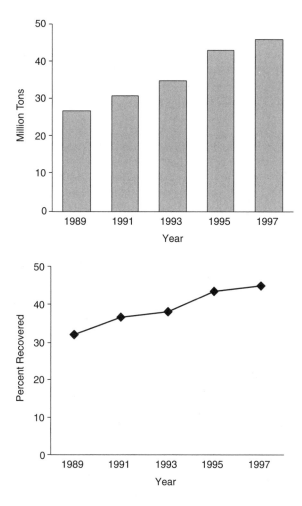

FIGURE 7-4  Paper recyling at U.S. mills.  SOURCE:  American Forest and Paper
Association (1998b).


tion, and municipal solid waste management[2] came to the forefront nationally.
These concerns extended across the industry's system of production and consumption as well as upstream to sources and management of the industry's raw

---

[2]These concerns were captured in an address to the 1992 Conference of the Technical Association
of the Pulp and Paper Industry by Peter Wrist, past president and chief executive officer of the Pulp
and Paper Research Institute, Canada, in which he said, "Today, our industry is under savage attack
by groups claiming that our operations are destroying irreplaceable natural treasures; poisoning the
oceans, lakes and rivers; burying out cities under piles of garbage; and threatening the health of the
public and the future of the world" (Wrist, 1992).

materials (i.e., trees and forests) and downstream to the industry's management of the effects of the nation's increasing paper consumption (Figure 7-1).

## CURRENT USE OF ENVIRONMENTAL METRICS

### Sustainable Forestry Practices

Environmental concerns related to acquiring wood from forests center around the method of harvesting, road placement, and local water quality. The industry uses a variety of metrics to track these concerns (Figure 7-5). Plantation forests can create impacts similar to those caused by agriculture (e.g., nonpoint sources of pollution from fertilizers). However, current attention is focused on the larger and longer-term effects of forests and plantations, such as the impairment of natural ecosystems, the health and diversity of species, and the economic resources of fisheries and recreation. The challenges for the pulp and paper industry are to define and refine sustainable forestry practice, adopt these practices, and measure progress toward them.

Forest sustainability has traditionally been assessed by the growth and yield of trees. The goal was to grow trees at least as fast as they are harvested to avoid creating a wood shortage (Boyce and Oliver, forthcoming). Forest practices aim to maximize growth rates and replant clear-cuts rapidly. In addition to clear-cuts, forests are also "thinned" occasionally, with the resulting wood usually being pulped. Trees harvested at a mature age are generally used for solid wood products and composites. Harvesting equipment has been developed to have minimal impact on the forest soil through compaction. Bio-solids and other

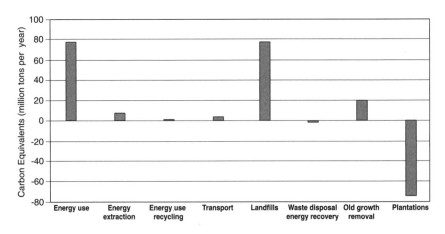

FIGURE 7-5 Greenhouse gas emissions from the paper cycle, carbon equivalents (million tons/yr). SOURCES: Grieg-Gran et al. (1997).

forms of fertilizers are sometimes applied to a forest to enhance growth rates. Replanting is generally done with one species of tree. Trees from genetic stocks that have strong, rapid growth rates are used, if possible. The metric used to measure growth rates is cubic meters per hectare per year. Sustainability is determined by comparing the metric with the harvest rate. If a forest is found to grow slower than its targeted rate, the management scheme may be modified or the harvest age prolonged. A related metric relates to reforestation, which is gauged by the fraction of the harvested area that is replanted.

According to Blum et al. (1997), managed regeneration forests and tree plantations affect

- soils and forest productivity, from harvesting or site preparation methods that can deplete nutrient levels over the long term;
- forest streams when activities such as harvesting, fertilizer and pesticide use, and road construction are performed without safeguards, such as adequate buffer strips along streams; and
- plant and animal habitat and species diversity—for example, resulting from the alteration of species composition and the physical structure of vegetation that, at a landscape scale, can reduce the available range of forest habitats.

The ecological effects of specific forestry management practices vary widely among different regions and depend on site conditions. Clear-cutting, for example, has potentially greater impacts in natural forests than in plantations or reforested marginal lands. Decisions to clear-cut natural forest are best guided by such factors as the natural forest disturbance regime (like fire-, wind-, or flood-adapted forest), where large-scale disturbances such as clear-cutting already occur. Other important factors are the characteristics of key forest species and the nature of the site. Alternative methods to clear-cutting, such as selective cutting, are generally less environmentally stressful but often lead to "high grading," where only the best-quality trees are harvested, leaving a low-quality stand. Selective cutting also leads to more frequent entries into a stand, increasing road and skid-trail use and thus more forest disturbances. In terms of land use, selective cutting also requires a larger land base than simple clear-cutting. Finally, the impacts of tree plantations depend on how and where plantations are established. Reforestation (including single-species plantations) on cleared and nonforested lands (such as marginal agricultural crop and pasture lands) is preferable to clear-cutting in many cases (Blum et al., 1997).

## Best Management Practices

In general, the basic level of environmental performance is gauged by the extent of adherence to best management practices (BMP) guidelines established

---

BOX 7-2
Tracking BMP Compliance as a Metric of Forestry Practices

Best management practices (BMP) are voluntary state guidelines intended to protect soil and water resources during forestry operations. Compliance to BMP can be used as a metric for forestry practices. Georgia-Pacific, for example, makes BMP compliance mandatory on all its nearly 6 million acres of forestlands. Forestry consultants are hired to conduct BMP audits on randomly selected tracts of land and use an independent panel of experts to review overall audit practices. In 1996, Georgia-Pacific achieved a 99.4 percent BMP compliance rate in a representative sample of forestry operation on company forestlands.

SOURCE: Georgia Pacific (1998).

---

by various states. An example of one company's effort to track and report on its BMP compliance is shown in Box 7-2. Most BMP relate to protecting water quality; some cover soil quality during forestry operations. BMP in the Pacific Northwest also address protection of wildlife habitat, natural communities, long-term soil productivity, and other forest values.

The challenge is to develop forestry management plans that produce quality wood while maintaining wildlife habitats and protecting water quality. Harvesting and replanting practices are changing to meet these new sustainability criteria. Replanting may involve multiple species, often with genetically selected species. A variety of harvesting techniques are used, including thinning, clearcutting, selective cutting, and shelterwood cutting. Harvesting is done in a way that leaves interfaces between forest structures (e.g., wooded areas, meadows, riparian zones). This provides habitats for species, like bats, that roost in mature trees but feed in open areas. Corridors in the cleared areas are left standing to provide passageways between wooded regions. Some trees are left in harvested areas and allowed to grow to old-growth stature. Care is taken to keep riparian zones forested to provide stream temperature control. Snags and other structures are left in the woods and in streams for habitat purposes.

Landscape management is used to meet BMP that address wildlife habitat protection. These practices are used to provide the diversity of forest structures (like clear-cuts and old growth) needed to maintain biodiversity (Boyce and McNab, 1994; Oliver, 1992). Research has shown that a mosaic of forest structures is required to support wildlife variety and maintain watershed health (Hunter, 1990). To maintain levels of biodiversity, the distribution of forest successional stages across a landscape has to remain constant. In order to achieve this, a portion of the landscape is harvested via clear-cut (stand initiation), a portion is harvested in an intermediate-growth phase (stem exclusion), and a portion is harvested in old-growth stage. If a different mix is desired, to support an endangered species for example, a new distribution may be used. The spatial and temporal scales over which these distributions apply will have an effect on the biodiversity of the region. Similarly, if the patchwork distribution of forest stages in riparian regions is maintained at historical (or known beneficial) levels, the watersheds will remain healthy. Metrics related to this practice are not well developed. New metrics are required to quantify the distribution of the forest successional stages and the distribution of forest structures in a region. Models need to be developed that relate new metrics to sustainability indicators, such as species variation and stream health. The metrics could then be applied at various spatial and temporal scales, depending on the forest sustainability objectives (McCarter et al., 1998).

Sustainable forestry (in the sense of habitat protection) is still experimental. The American Forest and Paper Association (1998c), an industry trade association, has adopted "Sustainability Forestry Principles and Implementation Guidelines." These guidelines designate broad objectives for member companies, leaving each company to design its own implementation to meet the goal. However, there is no specific performance standard. Companies track different things, such as landscape management, environmental auditing, company-specific BMP, special-area programs, logger training, and private landowner assistance programs (Blum et al., 1997), which makes assessment of compliance difficult. The challenge, and priority, in managing the environmental aspects of this part of the paper cycle is in developing a better understanding of the concerns of various stakeholder groups, many of whom have a wide variety of objectives.

One potentially interesting development in the industry, should there be a commitment to the Kyoto Protocol, would be addressing the benefits of carbon sequestration. Sequestration of carbon by trees as they grow would offset carbon dioxide production from paper mills and reduce the industry's production of greenhouse gases (Figure 7-5). An energy-related metric, such as tons of carbon sequestered per BTU used in the production process, could then be used to track the effect of carbon sequestration.

## Pulp and Paper Production

Figure 7-6 shows environmental metrics relevant to pulp and paper production. Emissions to air, water, and land are tracked and reported under the standard environmental regulations that cover the other industries discussed in this report. (See Chapter 4 for an overview of these regulations.) For purposes of comparison with other industries, pulp and paper metrics are discussed in terms of emissions; resource use; and reuse, recycle, and disposal.

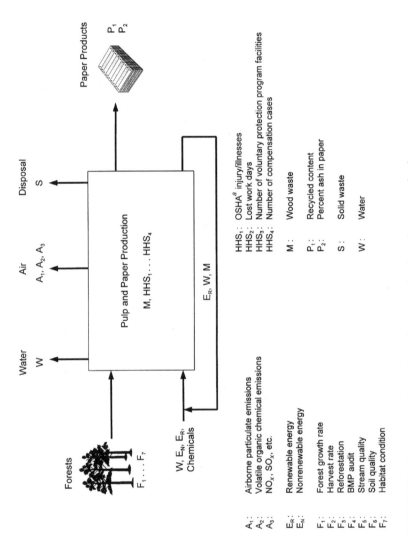

FIGURE 7-6 Environmental metrics in forestry and pulp and paper production.
[a]Occupational Safety and Health Administration

## Emissions

Air emissions are an important environmental concern in the pulp and paper industry. The primary emissions tracked include carbon monoxide, sulfur dioxide, nitrogen oxides, volatile organic compounds, and particulates. Some pulp and paper companies participate in the Environmental Protection Agency's Industrial Toxics Project (ITP) and are working to reduce emissions of 17 chemicals targeted by ITP. Sulfur emissions, which can cause odor (and public relations) problems, are also monitored.

The industry also tracks emissions to water through such metrics as total effluent flow per unit of production, total suspended solids, biochemical oxygen demand (BOD), chemical oxygen demand (COD) and color, and levels of chlorinated organics, such as dioxin and furans. Chlorinated organics are measured in terms of the weight of absorbable organic halides (AOX) per ton of pulp.

## Resource Use

Being the lowest-cost producer is a competitive advantage in an industry that is largely one of commodity production. Resource-related metrics include such things as raw materials and energy use, yield, and percentage of uptime help to facilitate reduction of production costs. Resource-related metrics, therefore, serve both business and environmental improvement goals.

Efficient use of wood as a raw material minimizes wood costs and thus increases the efficiency of a pulp and paper manufacturer's operation. While product quality concerns do not permit a 100 percent utilization rate, having as high a process yield as possible is a distinct business advantage with environmental benefits. Metrics associated with the use of wood as a raw material include percent yield of the processes and annual tons of wood waste disposed of in landfills. Production of recycled paper is resource related and is tracked by comparing nonwood fiber input with wood fiber input.

The pulp and paper industry is the third most energy-intensive industry in the United States. Metrics used to track energy usage are total renewable energy per ton of product and total nonrenewable energy per ton of product. The standard practice of using bark and wood waste and pulping liquor as (renewable) fuel eliminates more than 50 percent of the demand for nonrenewable fossil fuel in the industry as a whole—including in integrated pulp and paper mills (mills in which the paper-making operation is contiguous with the pulping operation) and nonintegrated mills (American Forest and Paper Association, 1994). Other steps that lead to energy conservation include reducing water usage, recovering and reclaiming high-level heat from digesters, improving insulation, and integrating systems that reclaim low-level heat.

Water is intensively used in paper making. A typical Kraft mill requires approximately 20,000 gallons per ton of pulp. Process innovations, such as high-consistency bleaching and hot-stock screening, require less water. Noncontact cooling, which segregates water from contamination, has also reduced the quantity of water used. Additional savings have accrued from internally recycling water using countercurrent washing and by reusing condensates, cooling and sealing waters, machine white water, and treated effluents. As a result, every gallon of water is reused an average of seven times in the process. The metric for water usage is gallons of water per ton of product.

One company in the industry (Weyerhaeuser Company, 1998) also counts among its resource reduction efforts an initiative to encourage car pooling to reduce commuter trips made by its employees. This and other efforts at reuse, recycling, and resource conservation are examples of activities in the industry that count toward its resource-related metrics.

### Reuse, Recycling, and Disposal

Most wood waste is burned as a fuel in boilers. However, bark, wood ash, sludge, and grit remain the chief solid wastes generated at paper mills. Solid waste is measured in terms of total solid waste disposal rates and through goals set for reducing wood waste sent to landfills.

Paper, the primary product of the industry, has been recycled for many years, but recently recycling has begun to receive greater attention. The recycled component of paper is often used as a marketing tool. For production purposes, recovery rates of paper that offset the use of wood fibers are tracked by the industry. In addition, because the integrity of the fiber is critical to the quality of products, the inherent paper-making value of the fiber is also important. There is, however, no reported metric for tracking degradation of fiber resulting from repeated recycling. One reason may be that the industry has several interconnected recycling and reuse options (Figure 7-7).

### Summary

Table 7-1 summarizes the environmental metrics used in forestry and paper production.

## CHALLENGES AND OPPORTUNITIES

### Sustainable Forestry

The greatest metrics challenge lies in defining sustainable forestry practices. There is a need to move beyond the traditional definition of growth and yield of trees and the current practice of compliance to BMP. Sustainability is a nascent area, and the development of more effective metrics will depend on improved

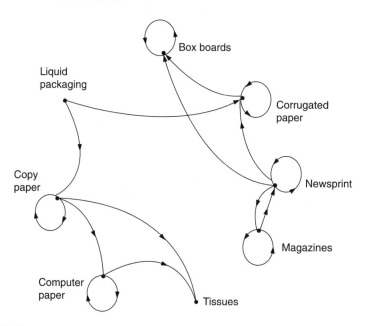

FIGURE 7-7   Recycling and intersectoral ecology in the paper industry.   SOURCE: Johnston (1997). .

understanding of ecosystems and the creation of suitable operational principles and practices.

## The Minimal Impact Pulp Mill

In an effort to provide a minimal impact vision for the industry, the American Forest and Paper Association (1994) has developed scenarios for a "Mill of the Future." The technological pathways to such as mill are shown in Figure 7-8 and involve improving the pulping and bleaching processes, since bleaching accounts for half of the effluent that comes from a Kraft mill. Current technology does not permit bleach effluent to be recycled. If sent through the recovery process, salts build up, leading to corrosion and scaling.

In the mill of the future, pulping will be done with energy-efficient digesters that promote extended delignification and with additives that enhance the pulping yield. Pulp will be bleached in closed or semiclosed bleach plants. These plants will require relatively small amounts of bleach chemicals with much of the spent liquors being recycled before they are sent through the Kraft recovery system. The bleach plants may contain separation devices to remove nonprocessed ele-

TABLE 7-1  Metrics Used in Forestry and Paper Production

| Forestry | Production | End Product (Paper) |
|---|---|---|
| *Resource Related* | | |
| Growth rate ($F_1$) | Percent yield | Recycled content ($P_1$) |
| Harvest rate ($F_2$) | Energy (renewable) ($E_R$) | |
| Reforestation ($F_3$) | Energy (nonrenewable) ($E_N$) | |
| BMP audit ($F_4$) | Water (W) | |
| | Wood waste (M) | |
| *Environmental Burden Related* | | |
| Stream quality ($F_5$) | Water—AOX, BOD, COD, TSS, | Percent ash in paper ($P_2$) |
| Soil quality ($F_6$) |    color, etc. ($W_1$–$W_N$) | |
| Habitat condition ($F_7$) | Air—Particulate, VOC, $NO_x$, $SO_x$, | |
| |    etc. ($A_1$ –$A_N$) | |
| | Solid waste (hazardous) ($S_1$) | |
| | Solid waste (nonhazardous) ($S_2$) | |
| | Environmental incidents | |
| |    Violation notices | |
| |    Citizen complaints | |
| | Permit excesses | |
| *Human Health and Safety* | | |
| OSHA[a] injury/illnesses ($HHS_1$) | | |
| Lost work days ($HHS_2$) | | |
| Percent of voluntary protection program facilities ($HHS_3$) | | |
| Percent of compensation cases ($HHS_4$) | | |

[a]Occupational Health and Safety Administration.

ments from the spent liquor before they are recycled. It is estimated that water usage in the bleach plant can be reduced to less than 4,000 gallons per ton of pulp (Histed et al., 1996). Effluent from the mill of the future will be significantly reduced with the level of chlorinated organics in bleach plant effluent declining accordingly. The pulping liquor in future mills will be gasified and burned in pulsed combustion furnaces. Such units can improve the energy efficiency of the combustion process by 40 percent while reducing $NO_x$ and $SO_x$ emissions. Finally, paper produced in the mill of the future will be dried with impulse presses. These presses lower the moisture content of paper by 20 percent, greatly reducing the steam required to dry the paper.

Metrics to assess the performance of the mill of the future will center on total use of resources. Because the mill of the future will use less water, a key measure will be improvement in the rate of reduction and reuse efforts. Another key metric will be energy use per unit of production. However, global climate change-related concerns may drastically affect the use and reporting of energy in the industry.

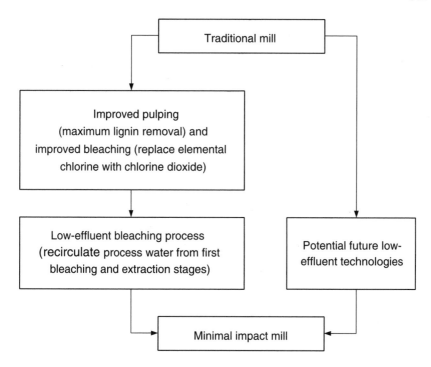

FIGURE 7-8 Pathways to a minimal impact mill. SOURCE: Adapted from Blum et al. (1997).

# REFERENCES

American Forest and Paper Association. 1994. Agenda 2020: A Technology Vision and Research Agenda for America's Forest, Wood, and Paper Industry. Available online at http://www.afandpa.org/Environmental/Agenda2020/index.html. [February 10, 1999]

American Forest and Paper Association. 1998a. Paper. Available online at www.afandpa.org/Paper/. [February 10, 1999]

American Forest and Paper Association. 1998b. U.S. Paper Industry's 50 Percent Paper Recovery Goal Progress Report. Available online at http://www.afandpa.org/recycling/Paper/programs.html#Progress. [February 10, 1999]

American Forest and Paper Association. 1998c. Sustainability Forestry Implementation Guidelines. Available online at http://www.afandpa.org/Forestry/guidelines.html. [February 10, 1999]

Blum, L., R.A. Denison, and J.E. Rushton. 1997. A life-cycle aproach to purchasing and using environmentally preferable paper. Journal of Industrial Ecology 1(3):15–46.

Boyce, S.G., and W.S. McNab. 1994. Management of forested landscapes. Journal of Forestry 92(1):27–32.

Boyce, S.G., and C.D. Oliver. Forthcoming. The history of research in forest ecology and siviculture. Chapter in Forest and Wildlife Science: A History, H.K. Steen, ed. Durham, N.C.: Forest Research Society.

Georgia-Pacific. 1998. Tracking BMP Compliance as a Metric of Forestry Practices. Available online at http://www.gp.com/enviro/wws_bpract.html. [February 10, 1999]

Grieg-Gran, M., S. Bass, J. Bishop, S. Roberts, N. Robins, R. Sandbrook, M. Bazett, V. Gadhvi, and S. Subak. 1997. Towards a sustainable paper cycle. Journal of Industrial Ecology 1(3):47–68.

Food and Agricultural Organization (FAO). 1995. Forest Products Yearbook. Rome: FAO.

Histed, J., N. McCubbin, and P.L. Gleadow. 1996. Bleach plant operations, equipment, and engineering: Water use and recycle. Pp. 643–667 in Pulp Bleaching: Principles and Practices, C.W. Dence, and D.W. Reeve, eds. Atlanta, Ga.: Technical Association of the Pulp and Paper Industry Press.

Hunter, M.L. 1990. Wildlife, Forests, and Forestry. Englewood Cliffs, N.J.: Prentice-Hall.

Johnston, R. 1997. A critique of life cycle analysis: Paper products. Pp. 225–233 in The Industrial Green Game, D.J. Richards, ed. Washington, D.C.: National Academy Press.

McCarter, J.M., J.S. Wilson, P.J. Baker, J.L. Moffett, and C.D. Oliver. 1998. Lanscape management through integration of existing tools and emerging technologies. Journal of Forestry 96(6):17–23.

Oliver, C.D. 1992. A landscape approach: Achieving and maintaining biodiversity and economic productivity. Journal of Forestry 90:20–25.

Simons, H.A. 1994. Water use and reduction in the pulp and paper industry. Vancouver, British Coumbia: NLK Consultants, Sandwell, Inc.

United States Department of Agriculture. 1993. Agricultural Statistics 1993. Washington, D.C.: U.S. Government Printing Office.

United States Environmental Protection Agency (USEPA). 1997a. The Pulp and Paper Industry, the Pulping Process, and Pollutant Release to the Environment. Fact Sheet. EPA-821-F-97-011. Washington, D.C.: USEPA.

United States Environmental Protection Agency (USEPA). 1997b. Pulp and Paper Rulemaking Actions. Washington, D.C.: USEPA.

Wernick, I.K., P.E. Waggoner, and J.H. Ausubel. 1997. Searching for leverage to conserve forest. Journal of Industrial Ecology 1(3):125–145.

Weyerhaeuser Company, 1998. Conserving natural resources. In 1997 Annual Environmental Performance Report. Available online at http://www.weyerhaeuser.com/environmnt/aepr97/default.asp. [February 10, 1999]

Wrist, P. 1992. Sustainable development and its implications for the forest products industry. TAPPI Journal 75(9):69–74.

# 8

# Summary of Metrics Used by the
# Four Industry Sectors

In the four previous chapters the committee examined the environmental metrics used or under development by four industry sectors. That information is summarized here to explore how much common ground exists among the sectors. Table 8-1 lists common or particularly utilitarian metrics. As in the sector chapters, no prioritization of these metrics is implied.

The meaning of most of the metrics is straightforward. "Supply chain" indicates corporations are looking at the environmental performance of their suppliers, using one or more metrics to do so. In the case of the pulp and paper sector, supply chain refers to the forestry part of the industry. "Emissions" indicates releases to land, water, and air; emissions to each medium are often tracked separately. "Percent of land preserved" refers to the use of land at corporate facilities; the metric is not yet precisely defined. Many of the metrics require normalization by some measure of business activity, such as energy use per unit of product or energy use per dollar of sales. "Sustainability" is undefined at present. The development of a suitable metric to address sustainability issues seems desirable.

As can be seen from Table 8-1, a number of metrics—emissions, energy use, materials use, water use, packaging, percent of recycled material used, and various measures of worker safety—are relatively common across all sectors, although the exact definitions may differ somewhat. Several sectors are dealing in some way with the question of sustainability, either by looking at the use of land, the emission of greenhouse gases, or by searching for an appropriate metric. A major difference among the sectors is in the area of product-related metrics, which are tracked extensively by the automotive and electronics sectors. For

TABLE 8-1  Environmental Metrics Used in the Four Industry Sectors

| Metric | Automotive | Chemical | Electronics | Pulp and Paper |
|---|---|---|---|---|
| *Supply Chain* | | | | |
| $M_1$ Supply chain | E | | E | |
| *Facility Centered* | | | | |
| $M_2$ Pollutant releases | C | C | C | C |
| $M_3$ Greenhouse gas emissions | C | C | C | |
| $M_4$ Material use | C | C | E | C |
| $M_5$ Percent recycled material | C | C | E | C |
| $M_6$ Energy use | C | C | C | C |
| $M_7$ Water use | C | C | C | C |
| $M_8$ Environmental incidence report | C | C | C | C |
| $M_9$ Lost workdays/injuries | C | C | C | C |
| $M_{10}$ Percent of land preserved | | E | | C |
| $M_{11}$ Packaging | C | C | C | C |
| *Product Centered* | | | | |
| $M_{12}$ Nongreenhouse gas emissions | C | | | |
| $M_{13}$ Greenhouse gas emissions | C | | | |
| $M_{14}$ Material use | | | | |
| $M_{15}$ Energy use | C | | C | |
| *Sustainability* | | | | |
| $M_{16}$ Sustainable forestry | | | | E |

NOTE: C = environmental metric in current use; E = emerging environmental metric.

industry sectors whose products are largely raw materials for others (i.e., chemical) or have essentially no impacts during the product-use phase (i.e., pulp and paper), product-related metrics do not appear to be useful.

## A GENERIC METRICS SET

If the metrics set identified in Table 8-1 were tentatively regarded as a suitable generic set for the four industrial sectors analyzed in Chapters 4–7, how well might the set serve other manufacturing industries?  To investigate this question, the committee briefly examined the suitability of the set for six additional sectors: agriculture, appliance manufacture, metal fabrication, mining, pharmaceuticals, and recycling facilities.  The following conclusions resulted from that analysis:

- The supply chain metrics ($M_1$, $M_5$) are generally suitable for all sectors. Even for a sector whose products are at the front of the supply chain (e.g., agriculture or mining), the metric could be applied to suppliers of equipment (e.g., combines, cranes) rather than to suppliers of raw materials.

- Emissions metrics ($M_2$, $M_3$) are applicable to all sectors.
- Resource metrics ($M_4$, $M_6$, $M_7$) are applicable to all sectors.
- Health and safety metrics ($M_8$, $M_9$) are applicable to all sectors.
- A land-use metric ($M_{10}$), while not yet well defined, applies equally well to these sectors as to the four sectors studied in detail by the committee.
- The product-packaging metric ($M_{11}$) is generally appropriate for all sectors, but packaging may not be required for some products (such as iron ore).
- Product-centered metrics ($M_{12}$–$M_{15}$) appear to be useful only for industrial sectors whose products have the potential for significant environmental impacts (e.g., appliance manufacturers).
- Until sustainability is better defined, the suitability of a sustainability metric ($M_{16}$) across sectors cannot be addressed.

Hence, it would appear that a provisional set of generic metrics might consist of:

$GM_1$—Materials use (normalized in some appropriate way)
$GM_2$—Water use (normalized in some appropriate way)
$GM_3$—Energy use (normalized in some appropriate way)
$GM_4$—Percent of recycled materials in products
$GM_5$—Percent of products that are leased
$GM_6$—Resource consumption and/or emissions during product use
$GM_7$—Average use of packaging
$GM_8$—Emissions from manufacturing (normalized in some appropriate way)
$GM_9$—Recordable health and safety incident rate
$GM_{10}$—A sustainability metric of some type

One could go on from this point to ask whether a generic environmental metrics set would be suitable not only for manufacturing industry sectors but also service industry sectors. To study that question, the committee imagined the use of such metrics in a set of hypothetical service-based businesses: a retail appliance store (A), a barber shop (B), a grocery store (G), a hospital (H), a lawyer's office (L), and a package delivery service (P). Table 8-2 assesses the suitability of the metrics set for these businesses. It appears from this table that the generic set, if carefully defined, is probably as useful to service industries as it may be to manufacturing industries.

TABLE 8-2  Fit of a Generic Metric Set for Hypothetical Service-Sector
Businesses

| Metric | A | B | G | H | L | P |
|--------|---|---|---|---|---|---|
| $GM_1$ | ? | S | ? | ? | S | S |
| $GM_2$ | U | S | ? | S | U | U |
| $GM_3$ | S | S | S | S | S | S |
| $GM_4$ | S | S | U | S | S | S |
| $GM_5$ | S | S | S | S | S | S |
| $GM_6$ | S | U | U | U | U | S |
| $GM_7$ | S | S | S | ? | ? | S |
| $GM_8$ | S | S | S | S | ? | S |
| $GM_9$ | S | S | S | S | S | S |
| $GM_{10}$ | ? | ? | ? | ? | ? | ? |

NOTE: S = suitable; U = unsuitable.

# PART III

# Current Status and Future Directions

# 9

# Observations, Trends, and Challenges

Following a careful review of the information and data available from the four industry sectors, several meetings that included presentations from outside experts, and a review of recent literature, the committee set about the process of analysis. The committee first assessed the current status of industry use of environmental metrics. Next, the committee used this information to identify trends likely to encourage the future development of metrics, as well as those that are currently having little effect. Finally, the committee sought to characterize the challenges that will have to be overcome for metrics to reach their full potential as an instrument for positive change. What follows is a summary of the committee's collective rendering of observations, trends, and challenges in the area of industrial environmental performance metrics.

## OBSERVATIONS

As the four industry studies demonstrate, U.S. companies have made significant progress over the past three decades in the development and application of environmental metrics. These metrics have proved instrumental in both documenting and driving progress toward environmental performance goals. As improvements in environmental performance are increasingly shaped by forces other than regulation, questions have arisen about what constitutes a goal and about the drivers associated with these goals.

Experience has shown that quantitative goals are a much more effective means of encouraging improvement than more vaguely defined statements, such as can often be found in corporate mission statements. Broad policy statements provide some framework for corporate action but are often insufficient to cata-lyze continual and substantive improvement. Employees and managers derive greater understanding and motivation from explicit performance standards. Clear, unambiguous metrics that can be used to monitor progress are an integral part of any effort to improve environmental performance. This is not meant to imply that concerns that are difficult to quantify (e.g., ecosystem health, biodiversity, and sustainability) should be disregarded, only that continuing efforts should be made to develop quantitative assessment criteria in these areas.

Observations from this study and the results of other relevant analyses (Delphi Group, 1998; KPMG, 1997; White and Zinkl, 1997) identify three general drivers for setting industrial environmental performance goals and developing associated metrics:

- the need to comply with regulatory mandates,
- the desire to achieve or strengthen competitive advantage, and
- the desire to improve corporate stewardship practices and reputation.

Effectively managing the environmental aspects of a major manufacturing operation with respect to all three drivers requires a broad range of metrics and a commitment to acquire what can sometimes be very specialized data.

## Compliance

The high degree of regulatory compliance achieved by U.S. industry over the past 30 years is laudable. Compliance is not only an accomplishment, it is also a prerequisite for moving toward the use of environmental performance as a tool to enhance competitiveness. All companies interested in long-term success readily acknowledge the squandering of money, time, and corporate image that results from continued violation of environmental regulations. Consequently, most organizations track their performance in areas such as accidents, spills, permit violations, and regulatory fines (KPMG, 1997; White and Zinkl, 1997). Many companies also make use of the information collected as part of government reporting requirements (e.g., related to emissions, waste generation) to assess their environmental performance. The Toxic Release Inventory (TRI), which obligates companies to report on the emission of over 700 different compounds, forms the basis of many environmental metrics used by industry. Regulatory schemes such as TRI are not without their problems but have provided a useful first step in the evolution of environmental metrics. Although some companies have recently begun to move beyond these types of measures to assess environmental performance, for most companies they are still the only metrics in use. Despite their limitations, compliance-based metrics have thus far provided the most consistent and useful industrial environmental performance information.

## Competitive Advantage

If industry is to continue expanding the frontiers of corporate environmental stewardship, the committee believes performance improvements must be linked to the profit-maximizing role of the firm. An increasing body of evidence supports the view that improved environmental performance and superior financial performance can go hand in hand (Blumberg et al., 1996; Cohen et al., 1995; Deutsch, 1998; Hart and Ahuja, 1996; White, 1995). However, there remains considerable debate over how strong this link is (Jaffe at al., 1995; Walley and Whitehead, 1994).

Although an ever-increasing number of firms claim to have enhanced their competitiveness through improvements in environmental performance, adherence to ecoefficiency principles is still far from the industry norm. Companies that have tried to improve the measurement and performance of their environmental stewardship efforts cite a number of benefits. Some of these are described below.

### Improved Process and Manufacturing Efficiency

Making more efficient use of materials and energy can significantly lower production costs in many industries (Allenby and Richards, 1994; Australian Environmental Protection Authority, 1997; Denison and Rushton, 1997; Frosch, 1995). As the price of scarce input materials rises and the cost of hazardous and nonhazardous waste disposal increases, better process metrics allow managers to analyze the cost-benefit trade-offs of improving overall materials efficiency. Energy is another industrial input for which costs are rarely trivial. Metrics that disaggregate energy use allow organizations to identify feasible cost savings options and protect against the volatility of the energy market.

### Increased Market Access

The committee notes that high-volume customers are increasingly using their market power to change the practices of their suppliers. Just as in the past these customers have pushed for higher levels of quality and reliability from their supply chain, some are now beginning to require environmentally friendly practices. German magazine publishers recently began to demand paper bleached by total-chlorine-free processes, while McDonalds and the U.S. government require specific levels of postconsumer recycled material in the paper products they buy. Commitment to environmental standards as a precondition to market access is becoming more common nationally and internationally. Before a firm can compete for a market, it must have access to it, a reality that may increasingly limit the potential customer base of companies whose environmental performance lags.

### Regulatory Advantage

Whether environmental regulations evolve toward ever-stricter command-and-control approaches or shift to more market- or risk-oriented schemes, companies that measure and manage their operations in an environmentally efficient manner will enjoy a distinct advantage. However, the committee believes that decisions to improve environmental performance, whether to meet a new regulatory standard or to free up emissions credits for resale in the future, require better information than is presently maintained by most companies.

### Impact on Shareholder Value

Analysts and shareholders both seem to agree that a discounting of corporate value is justified in the case of environmental laggards, although currently few seem willing to pay a premium for companies with a proven record of superior environmental performance (Gentry and Fernandez, 1996). The committee finds some evidence that this situation may change as metrics are devised that can reliably quantify certain financial risks (e.g., fines, cleanup, liability) incurred by conscientious organizations.

### Corporate Stewardship and Reputation

Maintenance and enhancement of corporate image together are another area in which environmental performance plays a significant and growing role (Robinson et al., 1998). While some voluntary environmental initiatives are prompted by competitive motivations (e.g., entering new markets, improving products), the committee observes that a significant portion are driven by the desire to improve a company's reputation. Growing societal awareness of environmental issues has placed increasing pressure on business to "do the right thing."

In today's business environment no company wants to be seen as falling behind, especially in an area with perceived ethical implications, such as the environment. The institution of a credible environmental performance ranking system, scored according to agreed-upon metrics, will generally promote competition between comparable firms. One need look no further than TRI to see the motivation it provided firms to lower targeted emissions. The pressure to reduce emissions was particularly keen among those with large quantities of TRI emissions, despite the fact that no regulatory intervention was imminent (Magretta, 1997).

Another less quantitative ranking is *Fortune* magazine's list of "America's Most Admired Companies" (Robinson et al., 1998). This survey rates companies based on responses from over 10,000 top executives, outside directors, and securities analysts. Companies are ranked in eight basic categories, one of which is "Corporate and Environmental Responsibility," and the eight scores are summed to provide an overall ranking. Of the 50 companies scoring highest in the environmental category in 1998, 32 ranked first overall within their respective industries, and only 2 companies ranked below the top third. Most executives sense that, while such information may not carry the same weight with stockholders or analysts as profit and loss data, corporate reputation is almost always a strong consideration in investment decisions.

Superior environmental performance can also have internal benefits in terms of enhanced corporate morale. Sensitivity to the environment among the general public has grown considerably in the last generation, and employees do not check their values at the door. Employees working for corporate environmental leaders often take great pride in their organization's accomplishments and are likely to make greater efforts to ensure that those feelings continue to be justified. Organizational pride and job satisfaction can help to attract and retain highly skilled personnel in today's competitive employment market.

Finally, the committee believes that more substantive environmental reporting can lead to improved relations with regulators and local communities, an advantage that may yield particular value by lowering the number of costly (in terms of both dollars and image) legal battles a company might face. Regulators continually state that they are much more likely to work toward cooperative solutions with companies that have demonstrated commitment to improving their environmental performance.

## TRENDS

The current trend in industry is toward greater public disclosure of information about waste generation and pollutant release (KPMG, 1997). While these and most other industrial environmental performance measures are responses to government reporting requirements, the committee finds evidence that some companies are beginning to move beyond compliance to report metrics associated with ecoefficiency. Most such efforts to date have understandably focused on meeting internal needs, but the increasing information demands of different external stakeholder groups have begun to influence the nature and format of environmental reporting.

Many metrics, particularly those related to compliance, have come to serve multiple purposes. Public stakeholders (e.g., local communities, environmental groups, regulatory agencies) use these measures as a proxy for assessing impact on public health and, to a lesser extent, on ecosystem health. Some commercial stakeholders (e.g., banks, insurers, investors) are also beginning to make use of environmental metrics. In the past these institutions and individuals were primarily interested in "negative" measures, such as those related to environmental accidents, violations, and liability. Recently, however, some stakeholders have begun to take notice of the more positive aspects of environmental performance (e.g., improved efficiency, lower risks; Deutsch, 1998).

Society's growing awareness of the environment is likely to continue to drive industry toward greater levels of disclosure. As public reporting has increased, there has been some movement toward more-standardized metrics (much of this facilitated by government requirements), but lack of comparability among companies is still a barrier to progress. Industry and nongovernmental organizations (NGOs) have both led recent efforts to establish standardized metrics. These attempts have focused principally on areas of broad application such as material- and energy-use efficiency; however, no consensus has yet emerged. (See Appendixes A–C for examples of metrics used across industry, within an industry sector, and within an individual company, respectively.) The lack of accepted metrics may contribute to the lack of interest and slow diffusion of best practices among small and medium-sized companies. At present, larger corporations are much more likely to investigate, develop, and use measures of environmental performance (Ehrenfeld and Howard, 1996; KPMG Denmark, 1997).

While larger companies have been at the forefront of implementing environmental metrics, in some industries the portion of the product life cycle under the direct influence of the firm contributes only a fraction of the overall environmental impact. Some companies are now beginning to think about the degree of influence they could reasonably exert over the supply chain, product use, and end-of-life disposition (Brown, 1998; Institute of Electrical and Electronics Engineers, 1997). One example is the program of product recycling and product reuse networks (somewhat akin to those in the auto sector) that some domestic electronics companies have begun to develop. (See Chapter 6.) Many of these efforts have begun to investigate final product disposition in both the United States and foreign markets (e.g., Japan, European Union), where product take-back legislation is now proliferating.

The expansion of producer accountability over more of the life cycle amounts to a first step beyond ecoefficiency toward sustainability. Although industry attempts to assess the sustainability of their products and operations are still at an early stage and currently involve only a few companies, public attention to the concept is likely to spur greater efforts in the future. One of the major challenges to the development of sustainability metrics is the vastness and complexity of the topic. Recent attempts to begin addressing global climate change, for example, highlight some of the difficulties in confronting such broad issues.

Society has shown growing interest in monitoring climate change, biodiversity, ecosystem health, and other indicators of sustainability, but the tools to do so are still crude. This problem is illustrated by attempts to enhance the utility of emissions data based solely on mass. Several efforts have recently been undertaken by industry, NGOs, and government (e.g., Imperial Chemical Industries' [ICI] Environmental Burden Approach, the Environmental Defense Fund's [EDF] Scorecard website, and the United States Environmental Protection Agency's [EPA] Hazard Ranking System, respectively) to rank the potential effects of environmental releases, not just tally the amount of material released.

These attempts to better quantify risk represent a useful step, but a great deal of concern exists over the level of uncertainty inherent in such ranking systems. Uncertainty is a serious obstacle to moving metrics into the realm of sustainability and comprehensive assessment of ecosystem impact.

## CHALLENGES

Despite the potential benefits of better metrics, there are some very real challenges standing in the way of efforts to improve the measurement, interpretation, and disclosure of environmental performance information. These barriers will have to be overcome by companies committed to continual improvement.

The limited availability and high cost of technology must be weighed early on when considering improved environmental metrics. Unless the technology exists to accurately and reliably quantify the desired parameter and at a reasonable cost, the acquisition of any data other than those required for compliance is unlikely. Other hurdles commonly present when implementing any new initiative, such as overcoming organizational inertia and developing corporate buy-in, will likewise need to be addressed if improved metrics are to realize their full potential.

While most of these industry-specific challenges will be dealt with by individual firms or facilities, the committee has identified a number of broader obstacles to the development and widespread use of improved industrial environmental metrics. Companies and industry associations are becoming increasingly interested in and capable of contributing solutions to national environmental problems. As the private sector continues to demonstrate a greater capacity to drive environmental improvement, the government's role must shift from that of a regulator to that of a facilitator. Nonetheless, because environmental quality is a "public good," a substantial, if declining, government role will still be required to effect change in these key areas.

### Improving Comparability

The four industry-sector studies underscore how the ability to compare performance, over time and across facilities, is key to improving the value of environmental metrics. To be useful as a management tool or as an element of public accountability, measures of environmental performance must be reliable, stable, and relevant. Unfortunately, many of the metrics in use by industry are not broadly comparable. Instead, they reflect the patchwork of disparate objectives from which they arose.

For those metrics dictated by mandatory reporting requirements, the underlying definitions and protocols have created some comparability. TRI provides one useful model, in this case of a governmental reporting system. Under TRI, over 20,000 U.S. manufacturing facilities must report in standardized fashion on their generation, release, and transfer of approximately 700 potentially toxic chemicals. Unlike other federal environmental databases, TRI uses consistent definitions of pollutants and facilities. This greatly simplifies the comparison of waste generation and pollutant release data over time as well as across plants, companies, and whole industries. This is not to say that TRI is an ideal cross-industry metric. Recent studies show that TRI may be tracking only 12 percent of emissions of hazardous substances under its purview (Axelrad, 1997), with the remaining fraction released from mobile or otherwise unregulated sources. Nevertheless, TRI is one of the only cross-industry metrics available.

In contrast to pollutant release data, measures of materials and energy use have not been guided by standardized reporting requirements. Instead, individual companies have developed their own metrics, usually as a element of cost control. For example, firms that purchase energy from off-site utilities can often rely on billing records to track electricity consumption. Materials-use efficiency is a metric that has proven quite useful to a number of companies for cost control, particularly those that use significant quantities of scarce or hazardous materials. Water is another resource that some companies have chosen to monitor.

Attempts have been made to develop consensus on a set of core principles and procedures for reporting environmental performance information. The Coalition for Environmentally Responsible Economies, for example, is assembling a broad-based group of companies, environmental organizations, social investors, and others in a global reporting initiative (White and Zinkl, 1997). Efforts like this, which increase the comparability of environmental metrics across industries, companies, and governments, have provided some benefit, but there is no widespread acceptance of any standardization scheme.

Another challenge to improving the comparability of metrics is that it is much easier to quantify an environmental burden than it is to assess the consequence of that burden. For example, a facility can tally the quantity of pollutants it releases to a water body relatively easily, but it is much more difficult to determine the impact of these pollutants on human, animal, or plant health. Equivalent burdens pose widely varying risks depending on many factors (e.g., the conditions of the receiving environment, its area, proximity to people, and the number of people exposed). The ecological impact of water discharges at the mouth of the Columbia River, for example, is quite different than the effect of the same discharges 100 miles upstream in critical salmon-spawning areas. Similarly, the health impacts of toxic air emissions in an urban area are quite different from the same releases occurring in unpopulated locations. Currently, environmental metrics do not reflect these differences. As we move toward metrics that are more closely calibrated to health and ecological impacts, it will be necessary to report releases in ways that reflect local conditions in a meaningful way. Devising data-reporting systems that are able to reflect local conditions while still maintaining a reasonable degree of comparability poses a significant challenge to standardization.

## Wider Dissemination of Best Practices

The environmental metrics used by the four industries examined by the committee reflect some of the best in U.S. manufacturing practice. The time is right to begin extending these successes to other parts of the U.S. economy, particularly small and medium-sized firms, to broaden the use of metrics across the product life cycle, and to promote the use of metrics beyond U.S. borders. This diffusion of environmental measurement practices has several key dimensions.

Over the past several years the largest manufacturers have been providing more detailed quantitative information on the environmental dimensions of their operations. As noted earlier, this trend in voluntary reporting has helped to boost the comparability of environmental metrics and win recognition for hard-earned improvements. Yet most small and many medium-sized companies have not felt the same pressures for public disclosure. As a result, their environmental metrics and goals rarely reach beyond regulatory compliance. One challenge will be to extend current best practices to the rest of industry.

Another concern frequently expressed is that as the environment comes to be seen as an avenue through which to seek competitive advantage, some companies will be reluctant to release publicly the details of their environmental operations. The need to protect privileged information will have to be addressed as corporate environmental reporting becomes more transparent and widespread. The risk that proprietary information can be extracted from strictly numerical data may be quite low. Both New Jersey and Massachusetts require the reporting of materials accounting data, including the amounts of hazardous chemicals brought on-site, generated on-site, and shipped off-site as either product or waste (Dorfman and Wise, 1997). Such reporting would seem to provide ample opportunity to reverse engineer processes and products. To guard against this possibility, both states provide relatively simple procedures for companies to protect their proprietary information. Yet only about 2 percent of companies have sought such protection (Hearne, 1996; New Jersey Department of Environmental Protection, 1995) indicating that in many cases even detailed data are of limited use without the expertise to correctly interpret them.

## Increasing Supply Chain and Life-Cycle Coverage

Supply chain relationships provide a valuable mechanism for spreading the use of environmental metrics beyond the largest firms. Through explicit requests for environmental performance information (and possibly through preferential contracting arrangements with firms that have better environmental measurement practices), some large firms and government procurement programs have encouraged suppliers to track and communicate environmental metrics. Companies with less purchasing power, of course, hold far less sway over their suppliers.

As long as the use of environmental metrics is restricted to company-owned facilities, vertically integrated companies will appear less environmentally responsible than competitors who outsource components. Yet apparent differences in corporate environmental performance speak to the differing boundaries of the life-cycle analysis, not the overall environmental burden associated with the full manufacturing process. This is especially problematic in industries where upstream suppliers operate in very different regulatory environments than the manufacturer of interest. Broadening the boundaries of analysis raises a number of questions. By what methods is an organization capable of influencing behavior up and down its supply chain? To what extent can it do so? And what are the reasons for doing so?

At present, organizations undertaking such efforts are generally characterized by strong executive commitment to environmental improvement. The U.S. Department of Defense (DOD) is one example. For many years DOD has maintained a sufficiently large procurement budget to allow it to impose a number of conditions on its suppliers, and some of these stipulations have related to environmental performance. Another example is the outdoor clothing company, Patagonia. Although not a manufacturer (it primarily assembles materials into finished products), Patagonia nonetheless insists that its suppliers purchase cotton raw materials only from businesses engaging in organic farming practices (Chouinard and Brown, 1997). Patagonia is an interesting case, since the company exerts considerable influence on not only its direct suppliers but also those vendors that do business with these suppliers. These examples (as well as several cited earlier) demonstrate the potential of organizations to begin leveraging their market power for environmental improvement. If such actions are to become more commonplace, however, suppliers will need much clearer definitions of what constitutes superior environmental performance.

At the other end of the life cycle, there are still relatively few environmental metrics describing the use and final disposition of products. The automobile, one of the most heavily regulated products from an environmental standpoint, offers some of the most familiar examples of product-oriented metrics. Fuel economy standards, tailpipe emissions limits, and gasoline specifications all center around the environmental burden of using, not making, cars. Given recent attention to climate change, it is particularly interesting to note that automobile use produces roughly four times more greenhouse gas emissions than automobile manufacturing. Energy efficiency standards and labels on consumer items help people buy products with lower energy costs and fewer environmental impacts. The concepts of extended producer responsibility, product take-back, and ecolabeling are further examples of approaches that push environmental performance far past the point of sale.

Extended producer responsibility is beginning to take on greater importance for global industries, particularly those that serve European markets. Sectors that have or may soon be impacted include the appliance, consumer electronics, and automotive industries. Considerations stemming from requirements such as product take-back add a new dimension to product design and manufacture, in both expected and unexpected ways. A semiconductor device manufacturer's environmental metrics, for instance, are not often related to the environmental issues facing downstream customers who incorporate the devices into finished electronic products. (See Chapter 6.) Producers of the finished goods, on the other hand, are experiencing significant pressures to provide for take-back of postconsumer products. The electronics industry, while still grappling with appropriate metrics, has responded with a number of new design practices. It is ironic that the quick turnover in computer technology that fuels the short lifespan of personal computers is driven by advances in semiconductors (and software), yet the burden of take-back, including reverse logistics, design for disassembly, and so on falls predominantly on the designers, developers, and assemblers of the finished products. Such inconsistencies are obstacles to the pursuit of improved metrics.

## Developing New Analytic Tools

Metrics epitomize the purposeful distillation of data into smaller, more manageable bits of information, thereby focusing attention on key issues. The danger is that pertinent facts may be lost or obscured in the process. As more industries assemble information on environmental performance, interest in tools to guide analysis and interpretation is growing.

Unlike many business metrics, measures of environmental performance are not readily translated into a single currency. Capital investments, projected revenues, labor costs, and even potential liabilities are routinely accounted for in the common language of dollars. Environmental metrics, however, are typically recorded in a hodgepodge of dissimilar units: pounds of hazardous wastes, parts per million of a chemical, number of oil spills, kilowatt-hours of electricity, and so on. The multiplicity of environmental metrics has led several companies and other stakeholders to propose weighting schemes useful for scoring across diverse dimensions of performance. (See Chapter 10 for an example of such a framework.)

So far, no single approach to weighting environmental metrics has received widespread support. Instead, some firms have begun developing customized tools for internal evaluation, while several nonindustrial organizations have begun applying their own criteria to publicly disclosed information. As noted earlier (Chapter 5), ICI, the U.K.-based multinational, has developed an Environmental Burden System combining different types of pollutants to evaluate potential environmental impacts in a handful of categories (e.g., ecotoxicity, aquatic oxygen demand, acidity, and hazardous air emissions).

These customized approaches may be useful for companies that endeavor to sharpen the focus of their environmental management. Unfortunately, because of the subjective (and often opaque) weightings applied, the composite indicators are not very meaningful to regulators, researchers, environmental groups, or customers. In addition to the disconnect with external stakeholders, most companies are reluctant to bear the expense of inventing a complex scoring scheme and then selling it to skeptical business units. While companies, regulators, and activists agree that focusing on the total quantity of a suite of chemicals (e.g., TRI) can be misleading, most fall back on such numbers as a readily available aggregate metric. As a result, companies that achieve large reductions in low-hazard chemicals appear to have had more success than companies that target more hazardous chemicals for reduction.

Community right-to-know initiatives and low-cost access to information via electronic media are opening up new ways to analyze and communicate environmental metrics globally. The new website created by the EDF links existing information on the generation and release of TRI chemicals from roughly 17,000 U.S. manufacturing facilities with a wealth of supplemental information (Environmental Defense Fund, 1999). A geographical interface puts selected facilities on maps at a national, regional, or neighborhood scale. With the click of a button, users of the site can rank these emissions by relative carcinogenicity, a quick means of weighting emissions of widely varying hazards.

Normalization is another issue that arises with virtually all environmental metrics. Take the example of a company whose production increases from one year to the next. All things being equal, the firm's environmental burden, in terms of energy and materials consumed, wastes generated, and pollutants released might rise proportionately. Which is the better metric, burdens per unit production or total burden per year? One would have held steady; the other increased. The answer is that it depends. To someone interested in ecoefficiency, the production-weighted metric is consistent with an overriding goal of less environmental impact per product. To a local community, total environmental loading may well be more important. There is no analytic solution to this basic divergence of goals.

The development of tools for evaluating environmental metrics, while in an early stage, will accelerate as businesses and stakeholders seek greater value from the information they already generate. For now, those who report and those who attempt to interpret environmental metrics must take great care to distinguish qualitative differences underlying quantitative metrics. That is, they must separate what is "big" from what is "important."

The links between emissions and environmental impacts are complicated by the difficulty of distinguishing individual from aggregate contributions (e.g., the effect of one more car on Los Angeles's smog problem). More profound is the fundamental difficulty tying cause to effect (e.g., higher concentrations of particulates in the air to increased incidence of lung disease). As a result, what firms measure is, at best, a proxy for what the public cares about. Further complicating matters is the considerable gap in the perceptions of environmental risk held by industry and the public (National Research Council, 1996).

## Addressing Emerging Environmental Issues

The four industry studies in this report demonstrate that many U.S. firms are using environmental metrics as an important management tool, especially in the continuing search for cost savings. Some have gone farther, articulating ecoefficiency goals to reduce the company's "environmental footprint" while also increasing its value. Several state and federal agencies are experimenting with alternative regulatory approaches for firms that pledge to improve and demonstrate their environmental stewardship. In addition, a full spectrum of organizations, community groups, and others is analyzing the utility of existing environmental metrics.

The industry studies also reveal an evolution in the use of environmental metrics as companies adopt environmental objectives in response to regulations, customer preferences, community pressures, and other forces. Since these external expectations change over time, companies must continue to update their internal metrics if they intend to meet future demands. Several emerging concerns are likely to influence the future choice and application of environmental metrics.

International attention to human effects on the global climate is a good illustration of the ways in which environmental concerns have motivated the development of goals and metrics for countries as well as companies. At present, no national legislation exists, and there is no consensus on how the 1997 Kyoto Protocol on climate change will affect U.S. industry. Nonetheless, each of the four industrial sectors has begun to estimate its emissions of greenhouse gases. For many manufacturing sectors, such as chemicals, this is primarily a question of energy use and $CO_2$ releases. The automobile and electronics industries share these concerns but focus even more on the effects of core products whose resource consumption during their use is greater than during their manufacture. The forest products industry, with relatively less dependence on fossil fuels but a much greater interest in the management of natural resources, will be influenced in even more complex and subtle ways. The forestry management side of the industry has begun work on metrics that quantify the moderating effect (i.e., carbon sequestration) of forest management on global climate change.

Other environmental issues on the radar screen of U.S. industry are likely to pose different types of measurement challenges. Concerns about the effects of chemicals on human health and development could significantly change the importance of whole classes of chemicals (e.g., endocrine disrupters) and modes of exposure. Efforts to gauge impacts on ecosystem health resulting from habitat changes, land management practices, changes in biodiversity, and other factors make the quantification of emissions to air and water look simple by comparison.

Even more challenging questions lay ahead as the public's demand for environmental protection broadens to encompass the economic and social dimensions of sustainable development. One example of this trend is the effort to assess traditional measures of pollution according to the race and economic status of those affected (a topic labeled "environmental justice"). Sustainability also has important implications for how industry views environmental metrics. Through the lens of sustainability, simply causing less environmental harm is not good enough. Goals are instead defined in environmental, economic, and social terms. Clearly, these concerns cannot be addressed solely with the environmental information at hand or by corporations alone. As with many emerging issues, sustainability creates new challenges for which existing environmental metrics fall short. Just as the shift from pollution control to ecoefficiency is forcing new definitions and measures of environmental performance, greater interest in climate change, ecosystem health, and sustainability are leading companies and their stakeholders to search for new yardsticks to track progress.

## Beyond Manufacturing

As we gain a better understanding of the environmental performance of industry, attention is turning to other contributors to environmental load. The committee notes that public-sector enterprises such as federal laboratories and defense facilities, while contributing significantly to environmental burdens, are only now beginning to come under the same sort of scrutiny that industry has become accustomed to. One example of this environmental double standard is the legacy of contamination present at some U.S. Department of Energy laboratories and military bases. The activities of large municipalities also have broad environmental implications. For cities and towns that operate major wastewater facilities, manage large vehicle fleets, and care for important public lands, environmental metrics will be as critical to their effective management as they are to corporations. In short, environmental metrics will be needed to meet increasing expectations for environmental quality across the entire spectrum of public and private institutions in the United States and internationally.

Finally, it should be pointed out that within industry environmental metrics are beginning to expand beyond the manufacturing sector. While outside the scope of this study, the activities of the service sector, including such disparate industries as retail sales, distribution services, airlines, energy services, and health care, have environmental implications that are just now being recognized (Graedel, 1997; Rejeski, 1997). Because the environmental impacts of service-sector firms result from activities dispersed across many locations, these companies generally escape the level of public or regulatory scrutiny that a major power plant or factory might receive. These impacts arise primarily through the logistics of deliveries to and from suppliers rather than from on-site activities. Ironically, such businesses often have tremendous opportunities to leverage environmental improvements through their purchasing power and direct link to customers. Nonetheless, it can be difficult to quantify either the environmental impacts of the operations of these companies or the potential to improve their environmental performance, largely because of the absence of useful environmental metrics.

## LOOKING AHEAD

Progress in devising and using industrial environmental performance metrics continues to be made. Many companies are in the process of experimenting with and developing new metrics. As these efforts proceed, near-term challenges will include improving the comparability and dissemination of best practices and developing methods for increasing the utility of current measures with techniques such as normalization and weighting. (See Chapter 10.) In the longer term, society's increasing interest in improving the environmental sustainability of human activities will require closer study of a host of poorly understood concepts and interactions. (See Chapter 11.) Better methods for reducing the impact of industrial activities on the environment will require better measures by which to gauge performance. Developing metrics that industry can use to improve its environmental performance will be a critical step in society's drive toward more sustainable practices.

## REFERENCES

Allenby, B.R, and D.J. Richards. 1994. The Greening of Industrial Ecosystems. Washington, D.C.: National Academy Press.

Australian Environmental Protection Authority (AEPA). 1997. Cleaner Production Case Studies. Publication 536. Melbourne, Victoria: AEPA.

Axelrad, D. 1997. How can TRI be used to provide environmental impact and health assessments? Paper presented at the Toxics Release Inventory and Right-to-Know Conference, September 8–10, Washington, D.C. Washington, D.C.: United States Environmental Protection Agency, Office of Policy, Planning, and Evaluation.

Blumberg, J., A. Korsvold, and G. Blum. 1996. Environmental Performance and Shareholder Value. Conches-Geneva, Switzerland: World Business Council for Sustainable Development.

Brown, M. 1998. Working with suppliers to improve environmental performance. Paper presented at National Academy of Engineering International Conference on Industrial Environmental Performance Metrics, November 1–4, 1998, Irvine, Calif.

Chouinard, Y., and M.S. Brown. 1997. Going organic: Converting Patagonia's cotton product line. Journal of Industrial Ecology 1(1):117–129.

Cohen, M., S. Fenn, and J. Naimon. 1995. Environmental and Financial Performance: Are They Related? Nashville, Tenn.: Investor Responsibility Research Center.

Delphi Group. 1998. A Business Guide: Environmental Performance and Competitive Advantage. Ontario: Queen's Printer for Ontario.

Denison, R., and J. Ruston. 1997. Recycling is not garbage. Technology Review 100(5):55–60.

Deutsch, C.H. 1998. Increasing evidence that green begets green. New York Times, July 19, Business Section, p. 7.

Dorfman, M.H., and M. Wise. 1997. Tracking Toxic Chemicals: The Value of Materials Accounting Data. New York: INFORM.

Ehrenfeld, J., and J. Howard. 1996. Setting environmental goals: The view from industry. Pp. 281–325 in Linking Science and Technology to Society's Environmental Goals. Washington, D.C.: National Academy Press.

Environmental Defense Fund. 1999. The Chemical Scorecard. Available online at http://www.scorecard.org. [January 21, 1999]

Frosch, R.A. 1995. Industrial ecology: Adapting technology for a sustainable world. Environment 37(10):16.

Gentry, B.S., and L.O. Fernandez. 1996. Valuing the Environment: How Fortune 500 CFOs and Analysts Measure Corporate Performance. United Nations Development Program (UNDP), Office of Development Studies, Working Paper Series. New York: UNDP.

Graedel, T.E. 1997. Life-cycle assessment in the service industries. Journal of Industrial Ecology 1(4):57–70.

Hart, S.A., and G. Ahuja. 1996. Does it pay to be green? An empirical examination of the relationship between emission reduction and firm performance. Business Strategy and the Environment 5:30.

Hearne, S.A. 1996. Tracking toxics: Chemical use and the public's "right-to-know." Environment 38(6):5–34.

Institute of Electrical and Electronics Engineers (IEEE). 1997. Proceedings of the 1997 IEEE International Symposium on Electronics and the Environment, May 5–7, 1997, San Francisco. Washington, D.C.: IEEE.

Jaffe, A.B., S.R. Peterson, P.R. Portney, and R.N. Stavins. 1995. Environmental regulation and the competitiveness of U.S. manufacturing: What does the evidence tell us? Journal of Economic Literature 33:132–163.

KPMG. 1997. 1996 International Survey of Environmental Reporting. Lund, Sweden: KPMG.

KPMG Denmark. 1997. The Environmental Challenge and Small and Medium-Sized Enterprises in Europe. The Hague: KPMG Environmental Consulting.

Magretta, J. 1997. Growth through global sustainability: An interview with CEO, Robert B. Shapiro. Harvard Business Review 75(1):78–83.

National Research Council. 1996. Understanding Risk: Informing Decisions in a Democratic Society. Washington, D.C.: National Academy Press.

New Jersey Department of Environmental Protection (NJDEP). 1995. Early Findings of the Pollution Prevention Program. Trenton, N.J.: NJDEP.

Rejeski, D. 1997. An incomplete picture. The Environmental Forum 14(5):26–34.

Robinson, E.G., A. Harrington, and J. Albany. 1998. America's most admired companies. Fortune 137:86–95.

Walley, N., and B. Whitehead. 1994. It's not easy being green. Harvard Business Review 72(3):46–50.

White, M. 1995. Corporate Environmental Performance and Shareholder Value. Charlottesville: University of Virginia.

White, A., and D. Zinkl. 1997. Green metrics: A status report on standardized corporate environmental reporting. Paper presented at the CERES Annual Conference, Philadelphia, September 30–October 1, 1997.

# 10
# A Hypothetical Model for Improving Aggregation and Presentation of Environmental Performance Metrics

Two characteristics of financial metrics that have made them very widely useful is that they can be applied nearly universally and they treat all financial issues thought to be of importance. The credibility of financial metrics has been hard won and is based on a foundation of generally accepted principles. Industrial environmental performance metrics, on the other hand, are nonstandardized and very sector dependent.

This chapter explores ways in which environmental performance metrics might be made more broadly useful through the selection of common metrics topics, normalization onto a common scale, and more effective presentation and aggregations. The generation and standardization of a set of metrics in this way are presented as a hypothetical model for aggregating proxies for environmental performance. Given the number of uses and users of metrics, it is unclear whether it would really be possible to collapse a wealth of metrics information into a handful of composite numbers. In addition, it would be difficult using such an approach to account for differences in geographic location or local circumstances or to characterize ecosystem-level impacts in any universally acceptable way. Nonetheless, this chapter illustrates the potential advantages of such a scheme over the present system of metrics used to gauge industrial environmental performance metrics.

## GUIDELINES FOR GENERIC METRICS

Unlike measures of financial performance, environmental metrics do not automatically lend themselves to a common unit. They tend to be recorded in such disparate units as pounds of waste generated, liters of water used, or hectares of forest harvested. Techniques exist for establishing common ground, however, if one relates performance on a percentage basis or some other uniform system.

A generic metric is sufficiently general to be used without regard to the industrial sector of the user, and a generic metrics set is a group of metrics that forms a reasonably complete picture of a corporation's environmental performance. A single such set, while it might provide useful guidance, is unlikely to serve all the metrics needs of all companies, or even all the metrics needs of a single company. However, a generic metrics set does contribute to the ability to assess competitive performance and is of considerable value in that regard.

Metrics come in broad classes, related to resources, environmental burdens, and human health and safety. While perhaps not strictly environmental, health and safety are often monitored by the same organization within a corporation that tracks environmental performance and are presented in corporate environmental reports, so it seems useful to consider them concurrently. Other metrics deal with products, suppliers, and broader environmental issues, such as sustainability.

In Chapter 8 the committee suggested a generic set of metrics divided into seven categories: resource metrics for manufacturing, products, and product packaging; environmental burden metrics; human health and safety metrics; supplier performance metrics; and sustainability metrics. The seven categories are described more fully below.

## Resource Metrics in Manufacturing

Resources are a natural focus for environmental metrics because of the potential, established to a greater or lesser degree for different resources, of eventual unavailability; the potential for some resource acquisition to entail unacceptable environmental impacts; and the feeling that a "single use and discard" approach to resources, especially nonrenewable resources, is inherently unsound.

For purposes of this example, three resource metrics, related to the use of materials, energy consumption, and water use, are considered. The units used to express these metrics are important. One cannot merely report mass of materials used, for example, because a yearly increase or decrease in the reported figure might simply reflect different production volumes. Rather, material use should be normalized to take production into account (e.g., pounds of material per pound of product rather than pounds of material per automobile manufacturing facility). One may also wish to consider the relative abundance of the materials.

The minimum value for a materials-use metric is obviously unity; that is, every molecule entering the facility leaves as part of a product. The maximum value is more difficult to specify, as it is almost certainly dependent on the industry sector. The smelter that extracts gold from ore is much less likely to avoid substantial residues than the goldsmith who crafts fine jewelry, for example. Performance is also heavily dependent upon suppliers. If a manufacturer receives metal produced by "near net shape" casting, for example, there will be less residue than if the same manufacturer begins with ingots.

Energy metrics relate both to the industrial processes employed and to the materials provided by suppliers. If minor assembly of components manufactured by others is the function of a facility, as is often the case in the electronics industry for example, the minimum energy may be close to zero.

Water metrics are directly related to the type of industrial processes in question. The minimum is obviously zero (no water use per unit weight of product); the maximum depends on the particular industry. In some industries, soft-drink manufacture for example, water is part of the product. In others it is used for cooling or as a constituent in chemical solutions. Water is often returned to its source cleaner than when it was removed, so an appropriate metric might be something like "gallons of degraded water discarded."

In addition to complexities resulting from the uniqueness of each industry sector, data availability may also limit the degree to which resource consumption-related metrics can be employed. Incoming materials and outgoing products may not be customarily available, for example. One might then consider related metrics, such as the cost of incoming materials divided by the cost of outgoing product, or the cost of incoming materials per unit of outgoing product. Such "second-level" metrics are much less desirable than "first-level" metrics because they are influenced by nonenvironmental factors such as financial negotiations with suppliers and customers, global financial oscillations, and so forth. The situation is equally unsatisfactory with regard to water and energy, for which geographically influenced supply conditions as well as financially related factors could blur the environmental evaluation.

Ideally, then, resource consumption-related metrics would be established by having industry consortia or other appropriate groups provide first-level maximum and minimum normalized values (for example, on a scale of 0 to 10). These scales will need to be periodically reviewed and updated to keep pace with improvements in knowledge and process technology. The performance of individual facilities or corporations could then be compared with those standards. Suppose, for example, that an industry were to establish standards as follows:

| | Minimum | Maximum |
|---|---|---|
| Materials (lb./$ of sales) | 3 | 10 |
| Water (gal./$ of sales) | 0 | 50 |
| Energy (BTU/$ of sales) | 1,000 | 2,500 |

Graphically, this would give

| Use | Rating | Use | Rating | Use | Rating |
|-----|--------|-----|--------|-----|--------|
| 3.0 | 10 | 0 | 10 | 1,000 | 10 |
| 6.5 | 5 | 25 | 5 | 1,750 | 5 |
| 10.0 | 0 | 50 | 0 | 2,500 | 0 |
| Materials ($M_1$) (lb./$ of sales) | | Water ($M_2$) (gal./$ of sales) | | Energy ($M_3$) (BTU/$ of sales) | |

Thus, a corporation using 6.5 lb. of material, 35 gal. of water, and 1,375 BTU per dollar of sales would have metrics of $M_1 = 5$, $M_2 = 3$, and $M_3 = 7.5$.

Materials use might also be expressed as yield, for example pounds of product per pound of input materials, or as pounds of input materials per unit of product. The choice should be made to maximize the potential for comparability among factories and companies, to allow long-term tracking, and to permit the easy acquisition of the necessary information.

The conservation of resources is strongly linked to several aspects of product design and customer interaction. Accordingly, metrics that measure these attributes are helpful indicators of environmental performance. The World Business Council for Sustainable Development (1996) has suggested several metrics of this type including

- the use of recycled materials,
- the use of renewable resources, and
- the provision of services rather than goods.

Metrics based on these measures may be strongly sector dependent, and industry consortia or other appropriate groups will need to establish more-descriptive definitions. One metric that seems particularly likely to be useful, however, is the percent of recycled materials in products (weight basis), or $M_4$. Both preconsumer and postconsumer recycling could be included. This metric can be transformed to the same 0–10 rating scale used for resource consumption metrics by dividing the percentages of recycled material by 10.

## Resource-Related Metrics for Products

A distinct group of resource-related metrics are those related to the use of products rather than their manufacture. The most common concern in this connection is with the use of energy, as with television sets, washing machines, or elevators. The unit of measure is generally the energy consumption of a single product over a given period of time, such as BTU/yr. Water and other ancillary inputs such as detergent or oil are other inputs with potential environmental significance that might be tracked.

As with other resource-related metrics, an industry group or corporation needs to establish energy consumption standards. For a television set, for example, the minimum consumption might be set as 0.5 kWh and the maximum at 1.0 kWh. The relative scoring scale is then

| Use | Rating |
|-----|--------|
| 0.5 | 10 |
| 0.75 | 5 |
| 1.0 | 0 |

In-service energy use ($M_5$) (kWh)

Thus, a television set drawing 0.75 kWh would have a metric of $M_5 = 5.0$. Other product-focused metrics could be related to water use (a washing machine), resource consumption (use of lubricants), or other topics of interest.

### Resource Metrics for Product Packaging

Packaging requirements differ greatly across industry sectors: It is obviously necessary to package a liquid chemical, but it is not so obvious that an automobile requires very much in the way of packaging. Within all sectors, however, the use of packaging is a useful measure of environmental performance, and a reasonable (though not technically comprehensive) metric is weight of packaging per dollar of sales. Thus, if an industry consortium set a packaging standard of 0.1 lb. of packaging per dollar of sales, the relative scoring scale becomes

| Use | Rating |
|-----|--------|
| 0 | 10 |
| 0.05 | 5 |
| 0.1 | 0 |

Packaging quantity (M6) (lb./$ of sales)

A corporation whose packaging use is 0.02 lb./$ would have a rating of $M_6 = 8$.

**Environmental Burden Metrics**

Compliance with regulatory requirements is a central feature of corporate environmental performance metrics, and such metrics are of obvious utility. However, simple reporting of tons of emissions gives little perspective on environmental performance. Normalization is needed. As with the water resource metric, the minimum level for the emissions metric is zero. The maximum level can be set by an industry consortium or some other appropriate group. These levels are then transformed to the 0–10 scale. For example, suppose that an industry-sector standard for maximum emissions were 0.5 pounds of emissions per dollar of sales. The rating scale would look like this:

$$\underline{\text{Emissions}} \qquad \underline{\text{Rating}}$$

$$
\begin{array}{ccc}
0 & \rule[0.5ex]{1.5em}{0.4pt} & 10 \\
0.25 & \rule[0.5ex]{1.5em}{0.4pt} & 5 \\
0.5 & \rule[0.5ex]{1.5em}{0.4pt} & 0 \\
\end{array}
$$

Emissions ($M_7$) (lb./$ of sales)

Emissions of 0.35 pounds per dollar of sales would have a metric of $M_7 = 3.0$.

**Human Health and Safety Metrics**

One or more metrics related to corporate health and safety performance is generally included in a company's environmental, health, and safety report. A common metric is the "recordable incidence rate," often expressed as the number of job-related injuries or illnesses per 100 employees. Zero is obviously the minimum for this metric. The maximum will be industry specific and should be established by an industry consortium or some other appropriate group. For example, suppose that an industry set the maximum rate at three incidents per 100 employees. Graphically, this gives

$$\underline{\text{Rate}} \qquad \underline{\text{Rating}}$$

$$
\begin{array}{ccc}
0 & \rule[0.5ex]{1.5em}{0.4pt} & 10 \\
1.5 & \rule[0.5ex]{1.5em}{0.4pt} & 5 \\
3.0 & \rule[0.5ex]{1.5em}{0.4pt} & 0 \\
\end{array}
$$

Health/safety ($M_8$) (incidences per 100 employees)

If a particular corporation had an incidence rate of 1.8, the metric value would be $M_8 = 4$. The use of several metrics is common in the human health and safety arena. Examples include annual worker fatalities and the number of lost work-days per employee.

### Supplier Performance Metrics

Corporations evaluate their suppliers on the basis of many characteristics, including financial stability, product quality, and manufacturing capacity. Some corporations are now also beginning to assess their suppliers' environmental performance. There are many variations on the practice, with some companies considering only whether their suppliers are in compliance with environmental regulations and others examining some of the aspects of the metrics described in Chapter 9. Using the 0–10 scale, a middle-of-the-road supplier might be rated $M_9 = 5$.

### Sustainability Metrics

Several industry sectors have suggested that one or more metrics dealing with sustainability might be useful. A potential metric of this type could convey information about the uses made of corporate lands. Such a metric would be very industry specific, since forest products companies or agricultural organizations obviously treat land much differently than do mining companies or electronics manufacturers. One way to define a land-use metric would be to divide corporate lands into three categories:

$\alpha$— land maintained in essentially a natural state;
$\beta$— land actively utilized for corporate purposes but with consideration given to habitat disruption, erosion, and related topics; and
$\chi$— land actively utilized for corporate purposes without significant consideration of long-term ecosystem impacts.

If $\alpha$, $\beta$, and $\chi$ are the fractions of corporate land determined (perhaps by an outside auditor) to fit the classifications as stated, we might arbitrarily assign a value of 1 to $\alpha$ lands, 0.5 to $\beta$ lands, and 0 to $\chi$ lands. This would result in the following scale:

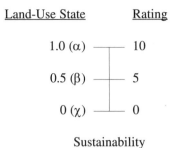

Sustainability

A weighted calculation is thus:

$$M = (1 \times \alpha) + (0.5 \times \beta) + (0 \times \chi)$$

For example, if a corporation has 10 percent $\alpha$ lands, 20 percent $\beta$ lands, and 70 percent $\chi$ lands, the calculation gives

$$M = (1 \times 0.1) + (0.5 \times 0.2) + (0 \times 0.7) = 0.2$$

This metric is transformed to the uniform scale by multiplying by 10:

$$M_{10} = 10[(1 \times \alpha) + (0.5 \times \beta) + (0 \times \chi)] = 2$$

Other potential metrics for sustainability can relate to specific impacts. A widely used metric is the emissions of particular chemicals, such as the chlorofluorocarbons and hydrochlorofluorocarbons that contribute to ozone depletion. Still other metrics might refer to habitat disruption, ambient noise generation, or some other impact of interest.

## USE OF WEIGHTING SYSTEMS

Some metrics may be more useful if weighting factors are applied. In the case of materials, for example, $M_1$, as initially defined, values a pound of platinum (a rare metal often in short supply) the same as a pound of silicon (an extremely common material). Without complicating the metric too much, one might take account of scarcity. One way to do this is to put all materials in three categories: those that are abundant, those whose supplies are moderately constrained, and those that are scarce. Arbitrarily, one can assign weighting factors of 1.0, 0.5, and 0, respectively, for these categories. The weighted version of metric $M_1$ then becomes

$$M_1 \text{ (weighted)} = M_1(1 \times \varepsilon) + (0.5 \times \lambda) + (0 \times \mu)$$

where $\varepsilon$, $\lambda$, and $\mu$ are the fractions (by weight) of abundant, constrained, and scarce materials, respectively, in the product under study.

A second metric for which weighting seems appropriate is that of emissions. Here, weighting factors can help distinguish materials that are nonhazardous, moderately hazardous, and highly hazardous. The weighted version of metric $M_7$ then becomes

$$M_7 \text{ (weighted)} = M_7 (1 \times \pi) + 0.5 \times \rho) + (0 \times \sigma)$$

where $\pi$, $\rho$, and $\sigma$ are the fractions (by weight) of nonhazardous, moderately hazardous, or highly hazardous emissions, respectively.

If weighting factors are to be used, it will be important to get community agreement on which material resources should be assigned to the scarcity categories and which emissions should be classed as hazardous. Such designations will have a significant impact on the usefulness of the metrics, and agreement on their use will be vital to their widespread acceptance.

## EXPANDED METRICS

In many cases a corporation may wish to develop detailed, as opposed to generic, metrics information. For example, a company may wish to express emissions to air, water, soil, and deep-well injection rather than to present a composite emissions figure. As before, minimum and maximum values would need to be established by an individual corporation or industry consortium, such as

|  | Minimum | Maximum |
|---|---|---|
| Air emissions (lb./$ of sales) | 0.1 | 1.0 |
| Water emissions (gal./$ of sales) | 3.0 | 7.0 |
| Soil emissions (lb./$ of sales) | 1.0 | 2.0 |
| Deep-well injection (gal./$ of sales) | 20.0 | 50.0 |

Graphically, this gives

| Use | Rating | Use | Rating | Use | Rating | Use | Rating |
|-----|--------|-----|--------|-----|--------|-----|--------|
| 0.1 | 10 | 3 | 10 | 1.0 | 10 | 20 | 10 |
| 0.6 | 5 | 5 | 5 | 1.5 | 5 | 35 | 5 |
| 1.0 | 0 | 7 | 0 | 2.0 | 0 | 50 | 0 |

| Air emissions ($M_{7a}$) (lb./$ of sales) | Water emissions ($M_{7b}$) (gal./$ of sales) | Soil emissions ($M_{7c}$) (lb./$ of sales) | Deep-well emissions ($M_{7d}$) (gal./$ of sales) |
|---|---|---|---|

Thus, for a corporation emitting 0.7 lb. of air emissions, 4 gallons of water emissions, 1.8 lb. of emissions to soil, and 29 gallons of deep-well injections, the expanded metrics would be $M_{7a} = 4$, $M_{7b} = 7.5$, $M_{7c} = 2$, and $M_{7d} = 7$.

## METRICS DISPLAYS AND METRICS AGGREGATION

The way in which metrics data are displayed and aggregated can increase their meaningfulness and utility. The danger of such approaches is that they can be simplistic and misleading; however, because of the potential payoff, they are worth exploring.

The simplest level of reporting conveys uniform values and trend information about individual metrics. The major benefit of having a common metric is that the uniform rating scale permits the environmental performance of corporations in the same industrial sector to be directly compared and the environmental performance of a single corporation to be tracked over time. It is not a significant problem that the scales will differ for different sectors. Such financial measures as debt equity and dividend payout ratios are routinely expected to be sector dependent, for instance.

To get an overview of a corporation's environmental performance, metrics can be displayed in a grouped chart. Suppose that 10 metrics have been selected and that all have been normalized on a scale of 0–10. We can now create a triangular metrics scoring display divided into 10 areas (one for each of the 10 metrics) (Figure 10-1). Within each area we express the rating score by a color, as follows:

| Red | Orange | Yellow | Cyan | Green |
|-----|--------|--------|------|-------|
| 0–2 | 2–4 | 4–6 | 6–8 | 8–10 |

Rating score

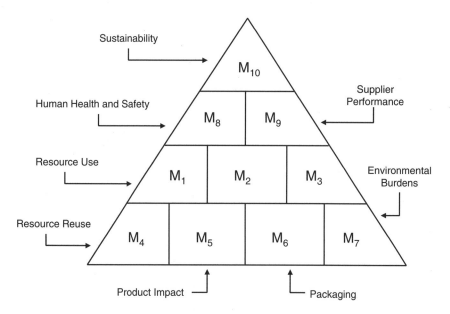

FIGURE 10-1 Triangular scoring display for metrics data.

A colored triangular display could show the metrics in a form that directs the user's attention to areas of highest and lowest performance, and it might be an efficient way to communicate the ensemble of industrial environmental metrics information applicable to an individual facility or a corporation as a whole. Such a display has the potential to become a uniform and readily understood presentation of corporate environmental performance, much as the U.S. Department of Agriculture's chart of daily minimum nutritional requirements efficiently presents the characteristics of foods or food products. A final though potentially controversial step might be to aggregate the results from a metrics set (Table 10-1) into a single indicator of corporate environmental performance.

The data in Table 10-1 can be shown in a series of triangle plots, as seen in Figure 10-2. The figure exploits the human ability to group patterns more quickly than numbers and colors more readily than shadings. Improvement in metrics 3, 5, 6, and 7 and a decline in metric 9 are quickly appreciated.

It is unclear whether a single environmental score for a corporation represents appropriate or inappropriate aggregation. Nevertheless, one could derive an overall environmental rating ($M_{sum}$) for a facility or corporation by adding the 10 metrics. In this model a perfect score would total 100. The result could be expressed as a trend over time (Table 10-2) or plotted as shown in Figure 10-3.

It is important to note that three distinct steps were involved in generating the composite metric shown in Figures 10-2 and 10-3. The first step was to select the classes of metrics. Three classes—related to resources, environmental burden, and health and safety—were chosen. For aggregation to work, agreement needs to be reached up front on which classes of metrics should be analyzed. The classes chosen will almost certainly be dependent on the industry sector. The second step is to select specific metrics within the chosen classes. Ideally, this step would also be independent of sector, though in some cases that may not be possible. The ideal number of metrics is not obvious, but more than 10 would likely be too confusing and expensive, while fewer than five seems unlikely to cover the necessary ground. This step will be harder to accomplish than the first. The third step is to assign minimum and maximum values for the chosen metrics; this is clearly sector dependent and is likely to be the most difficult to accomplish.

## SUMMARY

A set of corporate environmental metrics based on the model described in this chapter would have several uses. The aggregate rating would be universal, and the environmental performance of corporations in completely different industrial sectors could be compared at this level. At the level of individual metrics, corporations within the same industrial sector could be directly compared. At the intermediate level, that of metric classes, comparisons would be dependent or independent of industry sector, as a function of how the metric classes were established.

For a number of metrics groupings, arriving at a generic set of metrics appears to be a reasonable expectation. These groupings, however, measure corporate activity related to the environment, not corporate impacts upon the environment. For the foreseeable future, this is probably as much as can be expected, and if such metrics are widely used by corporations, the environment will benefit. As noted in the beginning of this chapter, the generic set of metrics and its possible use as a tool for comparison are but a model of what should be a relatively simple and straight forward process. Yet, as the four sector studies indicate, different users cast metrics in different forms, sometimes normalized to production, sometimes not, sometimes focused at the product or process level, other times at an entire facility, business unit, corporation, or nation.

This model is presented to show what may be possible and how such a scheme may be used. The steps outlined illustrate the need for consensus building on a set of simple, finite metrics; on the upper and lower limits used to rank performance; and on the relative positions of the various metrics within the triangle. This chapter is presented as food for thought, not as a final solution. It recognizes that there is a need for better metrics and better presentation of the information they convey. Arriving at a set of metrics that measures corporate

TABLE 10-1  Hypothetical Metrics Data, 1995–1998

|                | 1995 | 1996 | 1997 | 1998 |
|----------------|------|------|------|------|
| $M_1$          | 2.1  | 2.3  | 2.3  | 2.4  |
| $M_2$          | 4.6  | 4.5  | 5.7  | 5.8  |
| $M_3$          | 5.9  | 6.1  | 6.2  | 6.3  |
| $M_4$          | 5.1  | 5.2  | 5.4  | 6.1  |
| $M_5$          | 3.6  | 4.5  | 5.8  | 6.2  |
| $M_6$          | 3.5  | 3.7  | 4.1  | 4.3  |
| $M_7$          | 1.6  | 1.8  | 2.1  | 2.2  |
| $M_8$          | 7.2  | 7.6  | 7.9  | 8.1  |
| $M_9$          | 4.4  | 4.4  | 4.3  | 3.9  |
| $M_{10}$       | 8.4  | 8.6  | 8.7  | 8.8  |

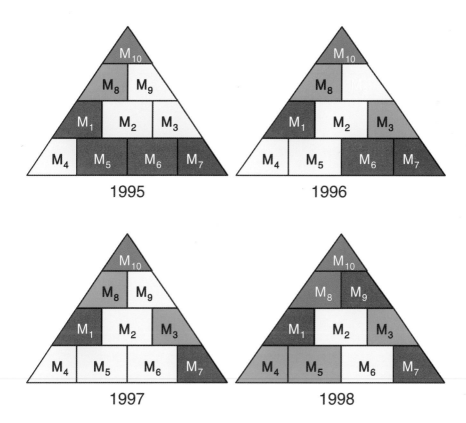

FIGURE 10-2  Triangular scoring display of hypothetical data, 1995–1998.

TABLE 10-2  Hypothetical Aggregate Metrics Scores, 1995–1998

| 1995 | 1996 | 1997 | 1998 |
| --- | --- | --- | --- |
| 46.4 | 48.7 | 52.5 | 54.1 |

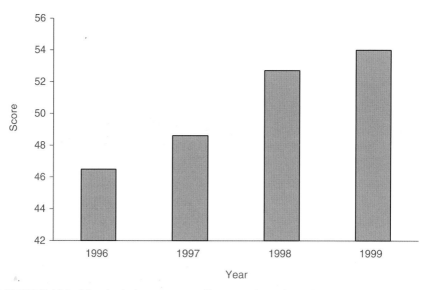

FIGURE 10-3  Hypothetical aggregate environmental metrics rating ($M_{sum}$), 1996–1999.

activity related to the environment is a difficult task.  Measuring corporate impacts on the environment, a much more difficult yet important topic, is the subject of the next chapter.

## REFERENCE

World Business Council for Sustainable Development (WBCSD).  1996.  Eco-efficient Leadership for Improved Economic and Environmental Performance.  Conches-Geneva, Switzerland: WBCSD.

# 11

# The Sustainable Enterprise Paradigm Shift

Current environmental practice and use of environmental metrics by a majority of U.S. industrial firms are best characterized as "cleaner production." As defined by the United Nations Environment Programme (1997), cleaner production is "the continuous improvement of industrial processes and products to reduce the uses of resources and energy; to prevent the pollution of air, water, and land; to reduce wastes at source; and to minimize risks to the human population and the environment." A few firms have embraced "ecoefficiency" as a standard of performance. Ecoefficiency makes the link between improved economic performance, higher resource efficiency, and lower environmental impact. It involves either "improving the productivity of energy and material inputs to reduce resource consumption and cut pollution per unit of output—in essence, making more and better products from the same amount of raw materials with less waste and fewer adverse environmental impacts" (World Resources Institute, 1998)— or using fewer raw materials or different, more environmentally benign materials. An even smaller number are beginning to assess the impacts of their activities within the context of sustainable development. This latter approach presents a daunting challenge, since it requires broadening the view of the relation between industry and nature to include community—a paradigm shift from "greening to sustaining" (Gladwin et al., 1995a; Hart, 1997).

In many ways the movement toward sustainability is a continuation of the "greening" shift that began in the 1970s. Since that time, companies have begun internalizing environmental costs in decision making. In many companies, environmental considerations are now integrated into core business functions such as research, development, distribution, provision of services, and product disposal.

This shift has been hard won and has required proof of environmental causes and effects and the development of new problem-solving techniques. It has been enabled by public policies and a range of technological innovations that reach well beyond environmental control or cleaner production. Innovations in information and communications technologies, in particular, have transformed production and management strategies throughout industry, changes that have brought unintended environmental improvements (Freeman, 1992).

The greening shift took about a generation to find mainstream acceptance. Time was needed to demonstrate the reliability of improved analytic principles and problem-solving techniques. Such a gestation period is not unusual. Kuhn (1962) has observed that new paradigms generally emerge without a full set of concrete rules or standards and often encounter considerable resistance that may require a generation or more to overcome. Sustainable development, first proposed in 1987, is complex and controversial, particularly when considered within the context of its social dimensions. A sustainable industrial enterprise incorporates and moves substantially beyond cleaner production and ecoefficiency. This chapter explores the idea of a sustainable development-oriented industrial enterprise and considers the challenges in developing associated performance metrics.

## The Call for Sustainable Development

The call for sustainable development is driven by observations that global environmental conditions are in decline and that significant environmental problems are deeply embedded in the socioeconomic fabric of all nations (United Nations Environment Programme, 1997). Presently, environmental decline is especially pronounced in the Asia-Pacific region, Latin America, the Caribbean, West Asia, and Africa (United Nations Environment Programme, 1997). Environmental trend data suggest a continued deterioration in the health of natural systems of these regions, taking the form of declining renewable resources, large-scale alterations of global biogeochemical cycles, and a threatened biological base (Brown et al., 1998; World Resources Institute, 1998). These environmental threats are cumulative and interactive, often arising from multiple causes, as shown in Figure 11-1.

Environmental deterioration is closely related to a variety of social trends.[1]

---

[1]These trends include increases in human population (projected to grow from about 6 billion at present to 9 to 10 billion by 2050) and associated issues of internal and cross-border migration and rapid urbanization (World Bank, 1998; World Health Organization, 1998); persistent deprivation, which is intrinsically linked to localized environmental decline and human disease (an estimated 3 billion people lack sanitation, 1.4 billion live in poverty, 1.3 billion lack clean water, 1 billion lack adequate shelter, 900 million adults are illiterate, 840 million people remain malnourished, and 35,000 young children in developing countries die each day from preventable disease [United Nations Children's Fund, 1998; United Nations Development Programme, 1997]); social disintegration resulting from the un- or underemployment (below a living wage) of 30 percent of the world's adult

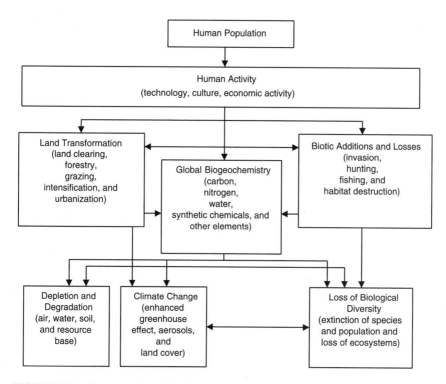

FIGURE 11-1  Conceptual model of human impacts on natural systems. Human effects, however manifested, have a systemic impact. Human and natural systems are increasingly coupled, especially at current and future scales of human activity. SOURCE: Allenby (1999). Copyright ©, 1998, Lucent Technologies. Used by permission.

Those in the business community who have responded to the sustainable development challenge cite 10 threats to ecosystem viability (Box 11-1). Technology and economic success can change the nature of these threats and their impacts on individuals and society. However, complexities associated with the interactions between human and natural systems make the quantification and management of such risks a daunting task (World Business Council for Sustainable Development, 1998).

---

work force, increasing income inequality (both among and between nations), rising crime, breakdown of families, and persistent gender bias (International Labour Organization, 1996; United Nations, 1995); and social inequality and growing resource scarcity (i.e., fisheries, forests, arable land, fuel wood, fresh water), which are increasingly combining to feed ethnic strife, political fragmentation, economic disruption, forced migration, and violent conflict (Homer-Dixon et al., 1993; Myers, 1993).

BOX 11-1
Ten Global Threats to Ecosystem Viability

1. *Loss of crop and grazing land* due to erosion, desertification, conversion of land to nonfarm uses, and other factors—about 20 million hectares a year.
2. *Depletion of the world's tropical forests,* leading to loss of resources, soil erosion, flooding, and loss of biodiversity—about 10 million hectares a year.
3. *Extinction of species,* principally from the global loss of habitat and the associated loss of generic diversity—over 1,000 plant and animal species each year.
4. *Rapid population growth.*
5. *Shortage of fresh-water resources.*
6. *Overfishing, habitat destruction, and pollution in the marine environment*—25 of the world's most valuable fisheries are already seriously depleted due to overfishing.
7. *Threats to human health* from mismanagement of pesticides and hazardous substances and from waterborne pathogens.
8. *Climate change,* probably related to the increasing concentration of greenhouse gases in the atmosphere.
9. *Acid rain* and, more generally, the effects of a complex mix of air pollutants on fisheries, forests, and crops.
10. *Pressures on energy resources,* including shortages of fuel wood.

SOURCE: World Business Council for Sustainable Development (1998).

The move toward the sustainable business enterprise presents an interesting mix of challenges and opportunities. The challenges lie in managing the long-term uncertainties inherent in resolving complex, coupled interactions between human and natural systems. Long-range analysis of trends in the efficient use of energy, materials, and land shows that it may be possible to decarbonize the global energy system and drastically reduce greenhouse gas emissions; that the material intensity of the economy can be reduced by leaner manufacturing, better product design, and smarter use of materials; and that it may be possible to increase the area of protected lands by reducing agricultural needs through the use of advanced farming techniques (Ausubel and Langford, 1997). In the short term there are opportunities to disseminate "best practices" in environmental management as well as environmentally friendly products and services. A cleaner environment has been achieved in concert with robust economic growth. Despite the successes of the U.S. system, there is considerable room for efficiency improvements. Even greater opportunities for global efficiency gains are to be tapped in less developed emerging economies.

## Sustainable Development and Enterprise-Level Metrics

Sustainable development was most influentially defined by the World Commission on Environment and Development (1987) as "development which meets the needs of the present without compromising the ability of future generations to meet their own needs." This original definition and subsequent refinements[2] remain controversial.

Nevertheless, metrics are emerging around three aspects—economic, environmental, and social—of sustainable development. Economic measures of corporate performance are tied to financial reporting and have evolved and matured over the past century. Environmental performance metrics, by contrast, began emerging only in the 1970s, and corporate environmental performance reporting appeared only in the last decade. Such reporting reflects a disparate and uncoordinated mix of metrics. Many firms provide qualitative descriptions of actions they have taken to improve corporate environmental performance, while some are beginning to provide an increasing amount of quantitative performance infor-

---

[2]These include to maximize simultaneously the biological system goals (genetic diversity, resilience, biological productivity), economic system goals (satisfaction of basic needs, enhancement of equity, increasing useful goods and services), and social system goals (cultural diversity, institutional sustainability, social justice, participation) (Barbier, 1987); to improve the quality of human life while living within the carrying capacity of supporting ecosystems (World Conservation Union, United Nations Environment Programme, and Worldwide Fund for Nature, 1991); that sustainability is a relationship between dynamic human economic systems and larger dynamic, but normally slower-changing, ecological systems, in which (1) human life can continue indefinitely, (2) human individuals can flourish, and (3) human cultures can develop but in which the effects of human activities remain within bounds so as not to destroy the diversity, complexity, and function of the ecological life support system (Costanza et al., 1991); that a sustainable society is one that can persist over generations, one that is far-seeing enough, flexible enough, and wise enough not to undermine either its physical or social systems of support (Meadows et al., 1992); that sustainability is an economic state where the demands placed upon the environment by people and commerce can be met without reducing the capacity of the environment to provide for future generations. It can also be expressed as " . . . leave the world better than you found it, take no more than you need, try not to harm life or the environment, make amends if you do" (Hawken, 1993); a vision of a life-sustaining earth whose people are committed to the achievement of a dignified, peaceful, and equitable existence. Under this view, a sustainable United States will have an economy that equitably provides opportunities for satisfying livelihoods and a safe, healthy, high quality of life for current and future generations. The nation will protect its environment, its natural resource base, and the functions and viability of natural systems on which all life depends (President's Council on Sustainable Development, 1994); and that sustainability is a participatory process that creates and pursues a vision of community that respects and makes prudent use of all of its resources—natural, human, human-created, social, cultural, scientific, etc. Sustainability seeks to ensure, to the degree possible, that present generations attain a high degree of economic security and can realize democracy and popular participation in control of their communities while maintaining the integrity of the ecological systems upon which all life and all production depend, while assuming responsibility to future generations to provide them with the wherewithal for their vision, hoping that they have the wisdom and intelligence to use what is provided in an appropriate manner (Viederman, 1994).

mation. Interest in so-called social reporting, first seen in the 1970s, has experienced some resurgence as companies have had to defend the working conditions and wages of their suppliers (e.g., Nike) or justify operating in countries with questionable human rights records (e.g., Shell). Such publicized incidents tarnish corporate image and alienate the customer base and, in extreme cases, can eventually lead to shareholder concerns about management policies and practices. Corporate sustainability accounting and reporting, while still somewhat nascent and exploratory, attempts to merge all three elements of sustainable development.

A review of sustainability-related economic, environmental, and social metrics by Fiskel et al. (forthcoming) reveals that there are currently well-established rules and standards for financial accounting and reporting. However, more sophisticated internal accounting metrics (e.g., activity-based accounting and economic value-added accounting) have helped reveal underlying drivers of economic performance and shareholder value (Blumberg et al., 1996).

If economic performance metrics are to evolve beyond merely accounting for profitability and cash flow, they need to quantify hidden costs associated with the utilization of materials, energy, capital, and human resources. They must also estimate uncertain future costs associated with external impacts of industrial production and consumption and lead to understanding of the costs and benefits incurred by various stakeholders (such as customers, employees, communities, and interest groups) across the life cycle of a product or process.

Environmental performance metrics are less well developed than financial metrics. Most are based in regulations and require companies to measure their output of wastes and emissions. While many companies track their material- and energy-use efficiencies, these are not commonly reported. Indeed, even within a single industry, there is great variety in what companies report as their environmental performance. Interest in standardization of environmental reporting is growing as individual companies experiment with ecoefficiency indicators. Consensus on an approach to measuring ecoefficiency could be an important prelude to any quantitative assessment of sustainability.

Social performance evaluations, perhaps the newest set of metrics to emerge in response to stakeholder pressure, are intended to track a firm's social accountability. They appear to fall into two categories: surveys of stakeholder responses on specific categories of performance (see, for example, those developed by the Body Shop [http://www.the-body-shop.com/news/index.html]) and attempts to evaluate social performance through community case studies (see, for example, British Petroleum [http://www.bp.com/_nav/commun/]). Metrics to assess social performance are embryonic and will require a great deal of development, improvement, and acceptance if they are to be truly integrated into business strategies and decision making.

Currently, there are no sustainability performance evaluations that attempt to integrate economic, environmental, and social measures. However, sustainability reports by Interface (1997), the U.S. carpet manufacturer, and Monsanto (http:// www.monsanto.com/monsanto/sustainability/), the newly reinvented life sciences company, indicate commitments to the concept of sustainable development and point to possible metrics.

## A Life Support Service Model of Sustainable Development

As envisioned by Gladwin et al. (1995b), sustainable development means protecting, maintaining, and restoring the integrity, resilience, and productivity of natural and social life support services (Figure 11-2). Ecosystem services, shown at the top, include purification of air and water, hydrologic regulation, waste treatment, soil fertility, pollination, pest control, seed and nutrient dispersal, biological diversity, climate and atmospheric chemistry regulation, culture, and aesthetics (Daily, 1997). By analogy, sustainability would also demand that a broad array of social system services that support citizen and industrial activity be protected and maintained, as displayed at the bottom of Figure 11-2.

This life support service model of sustainable development has been used to develop "working principles" for potential application at the level of the firm. For example, the ecologically sustainable enterprise, as envisioned by Gladwin and Krause (1996), would

- eliminate all harmful releases into the biosphere;
- use renewable resources such as forests, fisheries, and fresh water at rates less than or equal to their regeneration rates;
- preserve as much biodiversity as it appropriates;
- seek to restore ecosystems to the extent that it has damaged them;
- deplete nonrenewable resources such as oil at rates lower than the creation of renewable resource substitutes (e.g., solar) while providing equivalent services;
- continuously reduce risks and hazards;
- dematerialize, substituting information for matter; and
- redesign processes and products into cyclical material flows, thus closing all material loops.[3]

---

[3]These and other envisioned overarching rules of thumb are neither cast in stone nor do they take all aspects of systems into consideration. For example, the notion of using no more renewable resources than can be regenerated or depleting nonrenewable resources no more quickly than substitutes for them are found implies steady, even constant, rates of usage, which may not be realistic. It also hides the possibility of storing resources for which renewal is slow or for which substitutes need to be found and so may leave a false impression. For example, storage of resources may be an option as long as the "deficit" does not become too high.

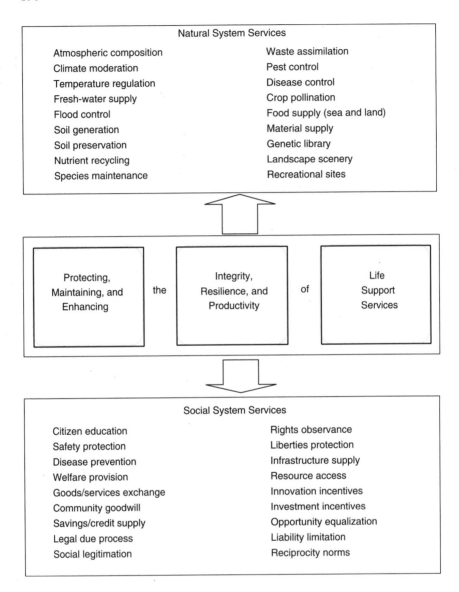

FIGURE 11-2 Sustainability as life support maintenance and enhancement.
SOURCE: Adapted from Gladwin (1992).

The socially sustainable enterprise, as envisioned by Gladwin et al. (1995b), would

- return to communities as much socially and economically as it gains from operating in them;
- meaningfully include stakeholders impacted by its activities in planning and decision-making processes;
- ensure no reduction in, and actively promote, the observance of political and civil rights in the domains where it operates;
- widely spread economic opportunities and help to reduce or eliminate unjustified inequalities;
- directly or indirectly ensure no net loss of human capital within its work forces and operating communities;
- cause no net loss of direct and indirect productive employment;
- adequately satisfy the vital needs of its employees and operating communities; and
- work to ensure the fulfillment of the basic needs of humanity prior to serving luxury wants.

These principles are beguiling in their simplicity, contestable on several fronts, and controversial in terms of the social orientation suggested. While offering some direction for how the shift to sustainability might be accomplished, they also lead to many complex questions. For example, the environmental implications of the design of a product such as an automobile or a service such as overnight delivery are often obscure. They depend on a variety of factors, including methods used to acquire raw materials, the environmental impacts of manufacturing wastes, how the product is used while in operation, and the product's final disposition. From a larger systems perspective, environmental impacts also depend on the number of automobiles on the road and their cumulative effect. From a social and economic perspective, transportation choices depend on cultural and social values.

## Toward Metrics of Sustainable Industrial Performance

Moving beyond pollution prevention and ecoefficiency to the sustainable development paradigm would require a profound transformation in the measurements and analysis used to gauge industrial performance. The set of transformations charted in Figure 11-3 involve large uncertainty, extraordinary detail, and dynamic complexity (i.e., nonlinear interactions between system components; significant time and space lags; complex feedback loops; unknown thresholds and irreversibilities; and multiple scales, resolutions, and rates of change). They also force the sustainability analyst into the realm of systemic interaction and aggregation, for conditions of sustainability reside in properties of "wholes,"

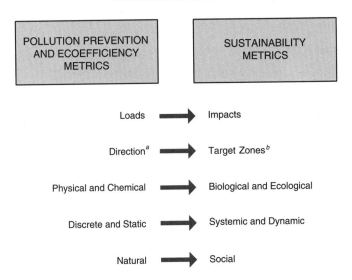

FIGURE 11-3   The transformation of pollution prevention and ecoefficiency metrics to sustainability metrics.
[a]Direction refers to trends such as reductions in energy, water, or materials use.
[b]Target zones refer to specific goals or ranges for targeted reduction or substitutions of resources.

which are something different or more than the sum of their parts. To determine, for example, whether the emissions from a factory are ecologically sustainable, one would first need to know, or at least be able to make reasonable estimates of, the conditions and assimilative capacities of all the ecosystems receiving those emissions, the character of all other disturbances flowing into those systems, the synergistic interactions among all the resulting stresses and biotic processes, and so on. This implies that impact assessments of a given product, process, or facility would need to be done with greater consideration given to these complexities and at broader temporal and spatial scales than those traditionally conducted by an individual firm.

It may be some time before the notion of sustainable industrial performance is translated into operationally measurable metrics. The five metrics described and included in Figure 11-3 provide a theoretical basis for achieving this goal. There are over 50 initiatives under way around the world to develop "sustainability rulers" for business. All of these efforts are struggling with the difficult challenges of complexity, comparability, credibility, and completeness (Ranganathan, 1998).

## From Loads to Impacts

The traditional environmental performance model advised industry to achieve improvements basically by reducing their loads, or footprints, on the environment (i.e., reducing material, energy, and service intensity; reducing or eliminating toxic dispersion; enhancing product durability and materials recyclability; etc.). The sustainability framework would demand, in addition, that all such behaviors be judged according to their actual impacts on the environment and society. Judgments about sustainable performance would depend on assessments of changes in the *states* of nature and society, something quite different from change in the *pressures* placed on nature and society. The sustainability analyst would also need to consider how gradual but persistent impacts can accumulate. "Preventing the slow, persistent, and cumulative degradation of natural systems resulting from human activity is the ultimate environmental challenge facing society. On a world scale, cumulative effects and sustainable development are inextricably linked, reflecting the mega-environmental problem and the mega-environmental solution, respectively" (Beanlands, 1995).

## From Direction to Target Zone

Pollution prevention mandates, product stewardship, energy and material efficiency, and cleaner production all contribute to lessening the environmental harms of industrial inputs and outputs. They fail, however, to provide guidance on how well industrial activities fit within the carrying capacities of local, regional, and global environments. Sustainability implies moving from metrics that measure environmental friendliness, consciousness, or greening to absolute benchmarks or "zones" of performance as determined by ecosystem and sociosystem health, the resilience and dynamic adaptability of life support systems, or social and ecological carrying capacities. The underlying calculus of eco-efficiency metrics is the efficient allocation of resources based on utility maximization according to market and price signals. The logic of sustainability would accept this goal but insist that it be constrained and defined by first assessing and ensuring that the scale and nature of human activities are sustainable.

## From Chemical and Physical to Biological

The industry case studies in this report reveal that the majority of currently employed metrics are concerned with material productivity, energy intensity, and toxic emissions. While these metrics will be of continued relevance and importance, the sustainability approach suggests that the primary threats to the ecosystem lie in habitat destruction (and the concomitant loss of biodiversity) and the exploitation of renewable resources (i.e., fresh water, fisheries, forests, wetlands, etc.) beyond their rates of regeneration. This pushes the performance analyst more deeply into the life sciences and emergent notions of ecosystem health (Costanza et al., 1992), ecological integrity (Woodley et al., 1993), and services. The sustainability analyst ultimately wishes to understand how a given industrial behavior affects the capacities of natural systems to maintain their vigor, organization, and resilience. Much of the requisite science and analytical technology to accomplish this has yet to be developed.

## From Discrete and Static to Systemic and Dynamic

Indicators of corporate environmental performance in use today are generally considered to be distinct and unconnected. The sustainability framework would require a holistic assessment, focusing on wholes rather than parts, of industrial performance in relation to impacted living systems. Such a methodology would examine interrelationships and feedback processes rather than linear cause-effect chains. Because dynamics and cyclicality are so fundamental in social and ecological systems, the sustainability approach would tend to focus on patterns of change in system structure rather than static snapshots. Sustainability metrics would build upon conventional metrics that have tended to focus on narrow or concentrated scales of space, time, and organizational complexity to inquire about impacts that are longer term and wider in geographic scale. The sustainability analyst would need to make connections across multiple spatial and temporal scales. Assessments would be sensitive to unknowns likely to arise from discontinuities (passing over unknown thresholds of disruption or irreversibility) and synergisms (problem interaction producing multiplicative rather than additive effects).

## From Natural to Social

Social progress is commonly acknowledged as one of the three pillars of sustainable development (along with ecological balance and economic progress), but it is typically downplayed or ignored in most business treatments of the topic. Problems such as poverty, gender bias, population growth, and environmental degradation are highly interdependent (Dasgupta, 1995) and often result in vicious downward spirals of ecological and economic decline (Durning, 1992).

Without gains in a variety of social factors, any gains in human progress derived from pollution prevention and ecoefficiency could be negated (Gladwin et al., 1995b). Sustainability metrics therefore must focus on industry's impacts on co-evolving social and natural systems. The sustainability analyst would need to simultaneously consider the consequences of industrial actions for ecological capital (i.e., renewable, cyclical, biological resources, processes, functions and services); material capital (i.e., nonrenewable or geological resources such as mineral ores, fossil fuels, fossil groundwater); human capital (i.e., people's knowledge, skills, health, nutrition, safety, security, motivation); and social capital (i.e., civil society, social cohesion, trust, reciprocity norms, equity, empowerment, freedom of association, other qualities that facilitate coordination and cooperation for mutual benefit). Environment-friendly developments such as cleaner automated factories or agricultural biotechnology would be assessed, for example, in terms of their employment consequences for factory workers and traditional farmers, respectively. The sustainability analyst is thus forced into the "morally thick" realm of social justice, appraising whether industrial activities shift costs or risks onto other human interests, today or tomorrow, without proper compensation.

## The Intellectual Journey Ahead

Moving from traditional notions of industrial environmental performance toward models and metrics of sustainable enterprise represents a long and difficult yet exciting and potentially rewarding journey. The journey has just begun. The scientific challenges of constructing and operationalizing impact-based metrics, geared to targets of sustainable systems, and gauging ecological and social consequences in a holistic and dynamic manner are extraordinary (Costanza et al., 1993). While the task of developing sustainability standards and associated metric systems will involve the scientific, governmental, corporate and nongovernmental communities, the responsibility for ensuring a sustainable world will fall largely on the shoulders of the world's businesses.

"Between the idea and the reality, between the conception and the creation, falls the shadow," said T.S. Eliot. The notions and metrics of sustainable enterprise currently lie in this shadow. The central task of corporate leaders moving into the next century is to bring them into the light.

## REFERENCES

Allenby, B.R. 1999. Industrial Ecology: Policy Framework and Implementation. Upper Saddle River, N.J: Prentice Hall.

Ausubel, J.H., and H.D. Langford. 1997. Technological Trajectories and the Human Environment. Washington, D.C.: National Academy Press.

Barbier, E. 1987. The concept of sustainable economic development. Environmental Conservation 14(2):101–110.

Beanlands, G. 1995. Cumulative effects and sustainable development. Pp. 77–88 in Defining and Measuring Sustainability: The Biogeophysical Foundations, M. Munasinghe and W. Shearer, eds. Washington, D.C.: World Bank.

Blumberg, J., A. Korsvold, and G. Blum. 1996. Environmental Performance and Shareholder Value. Conches-Geneva, Switzerland: World Business Council for Sustainable Development.

Brown, L.R., C. Flavin, H.F. French, J. Abramovitz, C. Bright, S. Dunn, G. Gardner, A. McGinn, J. Mitchell, M. Renner, D. Roodman, J. Tuxill, and L. Starke. 1998. State of the World 1998. New York: W.W. Norton.

Costanza, R., H.E. Daly, and J.A. Bartholomew. 1991. Goals, agenda and policy recommendations for ecological economics. Pp. 1–20 in Ecological Economics: The Science and Management of Sustainabilitly, R. Costanza, ed. New York: Columbia University Press.

Costanza, R., B.G. Norton, and B.D. Haskell, eds. 1992. Ecosystem Health: New Goals for Environmental Management. Washington, D.C.: Island Press.

Costanza, R., L. Wainger, C. Folke, and K.G. Maler. 1993. Modeling complex ecological-economic systems. Bioscience 43(8):545–555.

Dasgupta, P.S. 1995. Population, poverty and the local environment. Scientific American 272(2):40–45.

Daily, G.C. 1997. Nature's Services: Societal Dependence on Natural Ecosystems. Washington, D.C.: Island Press.

Durning, A.T. 1992. How Much Is Enough? London: Earthscan Publications.

Fiskel, J., J. McDaniel, and D. Spitzley. Forthcoming. Measuring product sustainability. Journal of Sustainable Product Design.

Freeman, C. 1992. Economics of Hope. Essays on Technical Change, Economic Growth and the Environment. London: Pinter Productions.

Gladwin, T.N. 1992. Building the Sustainable Corporation: Creating Environmental Sustainability and Corporate Advantage. Washington, D.C.: National Wildlife Federation.

Gladwin, T.N., J.J. Kennelly, and T.S. Krause. 1995a. Shifting paradigms for sustainable development: Implications for management theory and research. The Academy of Management Review 20(4):874–907.

Gladwin, T.N., T.S. Krause, and J.J. Kennelly. 1995b. Beyond ecoefficiency: Towards socially sustainable business. Sustainable Development 3:35–43.

Gladwin, T.N., and T.S. Krause. 1996. Envisioning the sustainable corporation. Pp. 9–24 in Stakeholder Negotiations: Exercises in Sustainable Development, A.R. Beckenstein, F.J. Long, M.B. Arnold, and T.N. Gladwin, eds. Chicago: Irwin.

Hart, S.L. 1997. Beyond greening: Strategies for a sustainable world. Harvard Business Review 75(1):66–76.

Hawken, P. 1993. The Ecology of Commerce: A Declaration of Sustainability. New York: Harper Business.

Homer-Dixon, T.F., J.H. Boutwell, and G.W. Rathjens. 1993. Environmental change and violent conflict. Scientific American 268(2):16–23.

Interface. 1997. Sustainability Report. Atlanta: Interface.

International Labour Organization (ILO). 1996. World Employment 1996/97: National Policies in a Global Context. Geneva: ILO.

Kuhn, T.S. 1962. The Structure of Scientific Revolutions. Chicago: University of Chicago Press.

Meadows, D.H., D.L. Meadows, and J. Randers. 1992. Beyond the Limits: Confronting Global Collapse—Envisioning a Sustainable Future. Post Mills, Vt.: Chelsea Green.

Myers, N. 1993. Ultimate Security: The Environmental Basis of Political Stability. New York: W. W. Norton.

President's Council on Sustainable Development (PCSD). 1994. A Vision for a Sustainable U.S. and Principles of Sustainable Development. Washington, D.C.: PCSD.

Ranganathan, J. 1998. Sustainability Rulers: Measuring Corporate Environmental and Social Performance. Washington, D.C.: World Resources Institute.

United Nations. 1995. World Summit for Social Development: The Copenhagen Declaration and Programme of Action. New York: United Nations.

United Nations Children's Fund. 1998. The State of the World's Children. New York: Oxford University Press.

United Nations Development Programme. 1997. Human Development Report 1997. New York: Oxford University Press.

United Nations Environment Programme. 1997. Global Environment Outlook. New York: Oxford University Press.

Viederman, S. 1994. The economics of sustainability: Challenges. Paper presented at the Workshop on the Economics of Sustainabilty, Fundacao Joaquim Nabuco, Recife, Brazil, Sept. 13–15.

Woodley, S., J. Kay, and C. Francis. 1993. Ecological Integrity and the Management of Ecosystems. Waterloo, Ontario: St. Lucie Press.

World Bank. 1998. World Development Indicators 1998. Washington, D.C.: World Bank.

World Business Council for Sustainable Development (WBCSD). 1998. Exploring Sustainable Development: WBCSD Global Scenarios 2000–2050. Conches-Geneva, Switzerland: WBCSD.

World Commission on Environment and Development. 1987. Our Common Future. Oxford: Oxford University Press.

World Conservation Union (IUCN), United Nations Environment Programme (UNEP), and Worldwide Fund for Nature (WWF). 1991. Caring for the Earth: A Strategy for Sustainable Living. Gland, Switzerland: IUCN, UNEP, and WWF.

World Health Organization (WHO). 1998. The World Health Report 1998. Geneva: WHO.

World Resources Institute. 1998. World Resources 1998–99: A Guide to the Global Environment. New York: Oxford University Press.

# PART IV

# Conclusions and Recommendations

# 12

# A Framework for Action

Over the past several decades, public concerns about risks to human and ecosystem health have driven individuals and organizations to act in a more environmentally conscious manner (Council on Environmental Quality, 1995; United States Environmental Protection Agency [EPA], 1992, 1998). The committee believes that this trend will continue as scientific understanding of environmental systems improves and society's demands for environmental improvement persist. As public attitudes continue to "raise the bar" with respect to environmental performance, each economic sector (e.g., agriculture, industry, municipalities) will choose the methods by which it will meet these challenges. If a particular sector's performance in the environmental arena is seen as inadequate, and if social pressure is maintained, government intervention—usually in the form of regulation—is likely. In the case of industry, past experiences with this prescriptive process have been viewed as particularly intrusive and inefficient. Limiting future regulatory actions will require skillfully harnessing performance improvements with both environmental and economic benefits while formulating innovative strategies to efficiently address environmental concerns that lack obvious connections to the bottom line. The committee is convinced that environmental performance metrics will play an important role in these efforts, providing a valuable tool to industry as it strives to do its part to lower human impacts on the environment.

The committee observes that concerns over compliance have driven the majority of environmental performance improvements. More recently, the private sector has discovered rewards in a more proactive approach to environmental stewardship. Companies and industry associations are becoming increasingly interested in, and capable of, contributing solutions to environmental challenges. As the private sector continues to demonstrate a greater capacity to drive environmental improvement, the government's role should begin to shift from that of a regulator to that of a "facilitator." In the future, partnerships between government, industry, and citizens' groups are likely to yield more creative and efficient solutions to environmental problems. This is not to say that government should abdicate all leadership on environmental issues. Environmental quality is a public good and, as such, the government must maintain a significant (if declining) role. While acknowledging this, the committee emphasizes that better results and greater efficiency have generally been obtained from companies that have voluntarily undertaken serious attempts at environmental improvement (Dow, 1996).

## ESTABLISHING A BASELINE: BEST PRACTICES

Analysis of the metrics in use by the more progressive organizations engaged in this study revealed a number of similarities in terms of the types of measurements tracked (Table 8-1). The committee feels that these environmental performance metrics represent a broadly accepted set of "best practices." However, best practices are far from common practices. Therefore, before suggesting ways of improving the current set of metrics, the committee wishes to provide some guidance to those organizations that have yet to establish a comprehensive framework of environmental metrics.

*RECOMMENDATION 1:* **Companies should investigate and implement to the greatest degree practicable environmental metrics representative of current best practices. Based on the four sector studies and the experience of its members, the committee urges firms to develop metrics in the 15 categories described in Table 12-1.**

These metrics categories fulfill several requirements. First, reliable and relatively unambiguous measurements may be derived in each of them based on present knowledge and technology. Second, many of the categories relate directly to core business concerns (e.g., cost cutting, improved efficiency), regulatory requirements, or the maintenance of good relationships with local communities or regulators. Lastly, the metrics require information that is already collected by most companies for regulatory compliance, inventory tracking, or waste management. That is not to say that resources will not be required to assemble the information into a usable form, but the means for obtaining many of these data is presently in place.

While the relevance of individual categories may vary by industry (the paper industry will, for instance, have greater interest in land-use metrics than the electronics industry), most have broad applicability. This recommendation thus provides guidance to the many companies that have yet to undertake a compre-

TABLE 12-1   Recommended Categories for Environmental Performance
Metrics in Manufacturing and Product Use

| Category | Brief Description and Examples |
|---|---|
| *Manufacturing Related* | |
| Pollutant releases | Includes:   *Air*—Some data collected to meet regulatory reporting requirements (i.e., Toxic Release Inventory [TRI]). Separated into hazardous/nonhazardous.[a] *Water*—Some data collected to meet reporting requirements. Similar to above. *Solid*—Some data collected to meet reporting requirements. Similar to above. |
| Materials use/efficiency | Separated into hazardous/nonhazardous.[a] |
| Energy use/efficiency | Broken down by resource (e.g., petroleum, natural gas, coal, renewable). Some companies have also begun to assess in terms of global warming potential (e.g., $CO_2$ equivalents). |
| Water use/efficiency | May track process water and cooling water separately. |
| Greenhouse gas emissions | Separated by gas (e.g., $CO_2$, $CH_4$, $N_2O$). Can be expressed in $CO_2$ warming potential equivalents. |
| Percent reuse/recycle/ disposal | Useful for assessing the use of individual process inputs as well as the final disposition of some intermediate products. |
| Packaging | Measured on either an absolute or per-product basis. |
| Land use | Separated into percent of land preserved, land developed, land restored, and inactive or abandoned developed land. |
| Environmental incidents | Classified by regulatory violations, fines, permit exceedances, accidents, etc. |
| Health and safety | Incidence of employee illness and injury and hours of training taken in safety, hazardous waste handling, etc. |
| *Product-Use Related* | |
| Pollutant releases | Includes:   *Air*—Separated into hazardous/nonhazardous[a] (e.g., emission standards for automobiles). *Water*—Same as above (e.g., output water quality of a washing machine or dishwasher). *Solid*—Same as above (e.g., toner cartridge for printer or copier). |
| Materials use/efficiency | Materials required for product use (e.g., detergent in cleaning appliances, fluids in automobiles). |
| Energy use/efficiency | Energy requirements for product use (e.g., corporate average fuel economy [CAFE] in auto industry, power use in electronic devices [EPA's Green Lights program], cooking efficiency [Electrolux]). |
| Water use/efficiency | Water requirements for product use (e.g., appliances, toilets). |
| Greenhouse gas emissions | Primarily a function of energy use. Can be expressed in terms of $CO_2$ (or $CO_2$ equivalents). |
| End-of-life disposition | Units or amounts of product reused, recycled, or disposed (may be further separated by method of disposal). |

[a]All references to a hazardous/nonhazardous distinction are made with respect to existing regulatory definitions in the United States.
NOTE: In many cases the usefulness of metrics will be enhanced by appropriate normalization (e.g., per unit product, per unit sales, per product use, per product lifetime).

hensive program of environmental performance measurement, and it provides a check for those companies with programs already in place.

At the level of the individual facility or firm, the committee strongly recommends that senior business leaders work with environmental managers to begin emphasizing the measurement and improvement of environmental performance in the areas cited in Table 12-1. The committee finds that environmental parameters have not been fully integrated into most business processes. These metrics should be built into approval procedures for new projects and products, and integrated into employee evaluation criteria.

The committee feels that, collectively, these categories of metrics represent an instructive and broad assessment of the present state of an organization's environmental performance and one that may be reasonably achieved. Although many further improvements to environmental performance measurement and reporting are needed, the widespread implementation of these metrics will be a significant and meaningful first step.

## GOALS FOR IMPROVING
## INDUSTRIAL ENVIRONMENTAL PERFORMANCE

To assist industry improve its stewardship of the environment, the committee has identified five goals and associated recommendations for enhancing the development and use of industrial environmental performance metrics:

- adopt quantitative environmental goals,
- improve methods of ranking and prioritizing environmental impacts,
- improve the comparability and standardization of metrics,
- expand the development and use of metrics, and
- develop metrics that keep pace with new understanding of sustainability.

In pursuing these goals, both government and the private sector have important roles to play. The unique ability of government to mobilize and assemble the wide range of individuals and groups with the required environmental expertise will be invaluable. Although government participation will be crucial to the success of many of the recommended actions, these steps should be based on scientific merit, not political expediency. Industry contributions to this "partnership" will be just as critical, as it is industrial firms that have the greatest expertise in evaluating and mitigating the environmental burdens imposed by their products and processes.

The committee's recommendations are based on an investigation of U.S. experience in improving environmental performance. While many of the proposed actions have broad, even global, applicability, they do reflect a U.S. perspective. Public attitudes toward the environment and regulatory frameworks will vary from country to country and may result in somewhat different approaches to improving industrial environmental performance.

## Goal 1: Adopt Quantitative Environmental Goals

The usefulness of a metric in driving environmental improvement is greatly enhanced by setting and tracking progress toward a quantitative goal. This is true at both the national and the corporate scales. Although in the case of emerging concepts such as sustainability numerical goals may not presently be feasible, setting firm goals with respect to many things that are now measured (e.g., mass of emissions, energy use, materials use) is an important exercise that has not yet been undertaken by many organizations.

At the level of the firm, management and employees need a strong grasp of exactly what is expected of them and the criteria by which they will be evaluated. Without this understanding, attention to environmental issues will be less focused and motivation will wane. A goal of 75 percent fewer TRI emissions or 50 percent more postconsumer recycle material, for example, represents objectives toward which an employee, department, or company can strive. Goals set at the national level should also be quantifiable. Such goals provide policy makers and government officials with clearly articulated objectives (e.g., 100 percent of cities must meet National Ambient Air Quality Standards) on which to focus throughout the sometimes convoluted political process. They also provide industry with a greater degree of certainty about the future regulatory environment, which is an asset to corporate planning activities.

*RECOMMENDATION 2:* **The U.S. government should strengthen its role in setting and reporting progress toward national environmental goals.**

The absence of clear priorities can lead to misallocation of resources. The federal government has a singular role to play in bringing together the technical expertise needed to prioritize the myriad environmental issues of national concern and to periodically update these assessments. The committee suggests that the process of developing quantitative national goals could be led effectively by the Council on Environmental Quality (CEQ) with heavy involvement from industry, citizen groups, and other stakeholders (e.g., EPA). Subsequently, a neutral body such as the National Research Council (NRC) could have a role reviewing these recommendations. Previous exercises of this type have been attempted (e.g., by the President's Council on Sustainable Development), but rarely have they articulated an explicit ranking of environmental priorities or established quantitative benchmarks. A more definitive rendering of environmental priorities will allow industry and other parts of society to direct their improvement efforts in a more efficient manner.

Quantitative goals, developed through a scientifically based consensus process involving government, industry, and environmental interest groups, have proved effective in addressing specific issues. Goals set as part of the Montreal Protocol, which led to the phasing out of compounds responsible for stratospheric ozone depletion, are a notable example of such cooperative efforts. Another example is the EPA Science Advisory Board's pioneering report, *Reducing Risk* (United States Environmental Protection Agency, 1990), and its predecessor, *Unfinished Business* (United States Environmental Protection Agency, 1987). The committee recommends that efforts of this kind be expanded and updated periodically to reflect improved knowledge and understanding of environmental systems.

> *RECOMMENDATION 3:* **Individual companies and industry sectors should set quantitative environmental goals and track and report their progress in meeting these goals. Individual companies and industry sectors should take the initiative in setting, tracking, and reporting on their progress in meeting quantitative environmental performance goals.**

The committee observes that while broad statements of policy such as those found in a corporate mission statement can provide some direction, they are often insufficient to catalyze continual and substantive environmental improvement. Companies that set quantitative environmental goals and commit to tracking and reporting progress toward them often realize rapid improvements in environmental performance. The committee also notes that while senior managers in most companies recognize a responsibility for safety and quality, far fewer have internalized a similar commitment to environmental concerns. If improvement in environmental performance is to be an effective and continuous process, all levels of corporate management must begin to bear some responsibility. Furthermore, while national environmental goals may not always be directly applicable to industry- or region-specific circumstances, related measures should be incorporated into corporate planning to the degree practicable.

The committee recognizes that important issues associated with ecosystem health, biodiversity, and sustainability still largely defy attempts at quantification. Eventually, alternative (i.e., nonquantitative) methods of assessment may reach an acceptable level of development to warrant widespread application. In the near term, however, the establishment of quantitative goals should be a primary objective.

### Goal 2: Improve Methods of Ranking and Prioritizing Environmental Impacts

Efforts must be undertaken to develop an acceptable system for prioritizing the issues of greatest environmental concern. Doing this will require moving from the measurement of environmental loads (e.g., air emissions, water emissions, resource use, land use) to the measurement of environmental impacts (e.g., human health impacts, ecosystem impacts). With such a system, goals and metrics can be established so that scarce public and private resources are directed toward reduction of environmental impacts in the most effective manner.

Companies often invest in a variety of voluntary environmental initiatives. Some of these efforts result in cost-effective lowering of environmental impacts and some do not. Such a framework would be valuable to industry and government as they continually seek to reassess and update their environmental goals. (See Recommendations 2 and 3.)

*RECOMMENDATION 4:* **Develop categorization systems to prioritize and target opportunities for reducing environmental impact. The U.S. government should facilitate a process with academia, industry, state agencies, and nongovernmental organizations to develop improved methods of ranking, categorizing, and prioritizing the relative impact of industrial environmental loads.**

This process should begin by focusing on human health risks and extend to issues of ecosystem health and long-term sustainability as knowledge and understanding of environmental systems evolve. Present knowledge may not allow for explicit numerical scoring of all impacts under all circumstances, but the committee feels that sufficient knowledge does exist to begin to prioritize categories of environmental loads (e.g., air emissions, water emissions, resource use, land use) relative to one another. The need for prioritization applies both within and across respective categories. Within the category of hazardous emissions, a number of efforts have been undertaken to rank substances with respect to their impacts on human health. Emissions may, therefore, provide a useful starting point. Data collected under TRI, as well as hazardous waste generation and disposal data collected under the Resource Conservation and Recovery Act (RCRA), may not fully represent a facility's environmental impact, but they do provide two of the only examples of consistent cross-industry metrics. Methodologies must also be designed to compare the relative environmental impacts of hazardous emissions against other categories of environmental loads, such as materials use, land use, or waste disposal.

The review of metrics used in the chemical sector (Chapter 5) notes the success of Imperial Chemical Industries' (ICI) implementation of its environmental burden system to identify and rank the impact of environmental releases within a number of categories (e.g., ecotoxicity, aquatic oxygen demand, hazardous emissions). While the ICI system does not establish a strict ranking of impacts across categories or prioritize the categories with respect to one another, it remains a useful tool. As society's knowledge and understanding of the complex interactions between environmental pollutants and long-term human or ecosystem health continue to become more sophisticated, efforts must be made to integrate this information into improved environmental metrics.

Because many of these topics are beyond the province of industrial research, government should assume a leadership role in bringing together industry leaders, academics, and public stakeholders to investigate metrics and goals that reflect cumulative and long-term environmental impacts. As a first step, CEQ, EPA, and the National Institutes of Health should jointly organize an effort to develop and introduce an agreed-upon system of ranking the substances already reported under TRI and RCRA with respect to environmental impact. Researchers and state and federal agencies have developed such systems in the past. All that may be needed, therefore, is to come to agreement on the most acceptable.

While the committee suggests this as a logical beginning given the ubiquity of TRI and RCRA, hazardous waste and air emissions account for only a portion of an industry's environmental load. Efforts must be made to begin to prioritize the broad range of potential categories of environmental loads. Given the diversity of manufacturing activities, such efforts would likely be most effective if performed at the industry-sector level. A program along the lines of the EPA Office of Compliance's sector notebook project, organized by EPA and involving experts from industry, academia, and public stakeholder groups, could begin to rank the environmental loads of greatest concern in various industrial sectors.

### Goal 3: Improve the Comparability or Standardization of Metrics

If environmental metrics are to become broadly useful across companies, industrial sectors, the investment community, and countries, great strides must be made in standardization. Corporate management desires greater uniformity in environmental metrics, both to support better internal decision making and to benchmark against competitors. Equally important, reporting standards will lend public credibility to industry efforts to improve environmental performance.

While industrial reporting of such things as waste generation and emissions has become more comparable, reporting in other areas such as energy use and natural resource consumption has lagged behind. Different companies and industries engage in a wide range of activities that can significantly complicate standardization efforts; however, sufficient similarities presently exist within several core ecoefficiency areas (e.g., materials efficiency, energy efficiency, water use) to begin devising comparable metrics. Established standards will simplify corporate efforts to develop and implement metrics while also providing credible evidence of improved environmental performance to external stakeholders.

*RECOMMENDATION 5:* **The U.S. government should facilitate a process of establishing consistent, standardized industrial environ-**

**mental metrics through the involvement of experts from industry, nongovernmental organizations, and federal agencies.**

The absence of standard metrics impedes benchmarking and reduces the value of public reporting of environmental performance. While reporting of standardized metrics is voluntary, a company claiming to have reduced its environmental burden should be judged by objective criteria (not unlike the formal definitions of terms like "low fat" or "nonfat" recently instituted in the food-processing industry). It is also possible that peer pressure may begin to push companies to report environmental data as more of their competitors choose to do so.

The committee is convinced that developing a set of standard metrics is absolutely critical to establishing a pattern of continual improvement in industrial environmental stewardship. While government participation is prudent in any process that seeks to set national standards, industry should play an integral role in the development, implementation, and promotion of standardized environmental performance metrics. This effort includes not only the establishment of standard measures but also the identification of units of normalization that achieve the broadest applicability (e.g., units per dollar of operating income or per-dollar value to society). Individual industrial sectors must help determine the relative value and practicality of specific environmental metrics while also working to encourage the adoption of standardized metrics across the corporate world through avenues such as the supply chain and trade associations.

The committee recognizes a number of commendable efforts to define and establish a simple, robust set of environmental metrics. However, these attempts have been largely uncoordinated. Some organizations with significant market power have successfully insisted that their suppliers meet and report on environmentally derived standards. Industrial associations have likewise played a role in the promotion of standard practices, but such efforts have often been hampered by the wide range of constituents these organizations must consult before advocating new positions. Government and nongovernmental organizations have also contributed to broadening the knowledge base. The World Business Council for Sustainable Development, for example, has made efforts to establish eco-efficiency metrics and has provided critiques of current environmental reporting methods. EPA has also provided valuable information through its Pollution Prevention and Green Accounting initiatives. Another recent development is the Global Reporting Initiative, an effort to identify and design standard metrics being conducted by the Coalition for Economically Responsible Economies with the participation of organizations ranging from General Motors to the United Nations Environment Programme.

The committee notes that CEQ and the NRC are both capable of providing leadership in efforts to define and standardize environmental metrics. Industry expertise and the EPA Science Advisory Board should be engaged in all phases of this work. The National Institute of Standards and Technology, which has a long history of conducting standardization programs, should be heavily involved in the effort.

*RECOMMENDATION 6:* **The U.S. government should promote standardized industrial environmental performance metrics in international forums.**

The world is becoming increasingly interconnected, and the environmental performance of a U.S. corporation is often judged by much more than local or national standards. The same applies to foreign multinationals operating in the United States. Some method of comparing environmental performance across countries is required.

In today's global economy, corporate operations are not limited by national boundaries but depend on extensive global supplier chains and distribution networks. Establishing international industrial environmental performance standards will not be easy. However, establishing a framework for objectively measuring and comparing different nations' environmental performance will provide incentive for organizations to improve their operations in countries with varied commitments to environmental protection and enforcement.

The committee suggests that U.S. aid and trade agencies might begin this process through their technology transfer and exchange programs. Another course of action might involve collaboration among the Office of the U.S. Trade Representative, the U.S. Agency for International Development, and the U.S. Department of Commerce (DOC)—coordinated by the U.S. Department of State—to initiate an international process to bolster the comparability of environmental metrics. Accepted standard environmental measures of industrial processes, products, and marketing might also be promoted for use by international organizations. One excellent example of such interaction is the leadership provided by the U.S. government in the development of the Organization of Economic Cooperation and Development's Pollutant Release and Transfer Registers. The 1997 Kyoto Protocol on climate change provides a superb opportunity to test the international community's ability to develop standardized environmental metrics.

## Goal 4: Expand the Development and Use of Metrics

The four industry studies in this report reveal that some companies, primarily large manufacturers, have made great strides in measuring and improving their environmental performance. The time is right to expand the use of environmental performance metrics over more of the product life cycle and to disseminate knowledge of best practices to a wider audience.

In recent years some of the largest manufacturers have been providing more detailed quantitative information on the environmental dimensions of their operations. Environmental measures in the manufacturing stage are important, but attention must now shift to other life-cycle areas such as raw material extraction and refining; product use; and product disposal, reuse, and recycling. Some industries have already begun to use metrics in these areas for either competitive or regulatory reasons. Many paper and forestry products companies maintain their own raw material reservoirs and thus have a vested interest in metrics related to rates of harvest and sustainable land management practices. In the case of automobile manufacturers, a number of product-use metrics such as vehicle emissions and average fuel economy are closely tracked as a matter of regulatory compliance. Growing interest in sustainability has led to the consideration of product take-back legislation in many countries and resulted in increasing emphasis on improved metrics to assess product end-of-life characteristics. As these examples indicate, life-cycle metrics exist, but they are still far from commonplace.

One challenge is encouraging the development of metrics within the manufacturer's supply chain. Even in cases where suppliers can be encouraged to measure, collect, and report environmental information, compiling comparable data from a range of suppliers can be difficult. There are also limits to the depth to which life-cycle attributes (and the supply chain) can reasonably be investigated. Despite these challenges, the potential environmental benefits of viewing the product life cycle more holistically demand that the corporate boundaries of environmental performance metrics be enlarged.

*RECOMMENDATION 7:* **Industry should integrate environmental performance metrics more fully throughout the product life cycle.**

Few companies or industries control their product from cradle to grave, and many exercise direct influence over only a fraction of product life cycle. Industrial executives, managers, and engineers should begin to extend the application of environmental performance metrics both up (e.g., to account for product use and end of life) and down (e.g., to account for raw materials acquisition and processing) the supply chain.

The committee recommends that companies and industrial sectors take the lead in more fully assessing the life-cycle impacts of their products. This process may include conducting surveys or studies in partnership with suppliers to determine which metrics are most useful and feasible. In large corporations, some of which contract with hundreds or even thousands of suppliers, this task may appear overwhelming. However, these larger companies also maintain substantial leverage over their suppliers in terms of dictating product characteristics. Attempts should be made to begin investigating the potential for robust supply chain metrics, even if only in connection with a lesser number of larger vendors. The committee believes that an invaluable contribution to this effort would be the establishment of a consistent, cost-effective methodology for providing estimates of relative life-cycle impacts. Comprehensive life-cycle analyses often become intractable as the level of investigation becomes ever deeper and more detailed. A system of quick yet consistent assessment needs to be developed, standardized, and promoted. A systems approach will be required, if suppliers, manufacturers, consumers, and those responsible for the end of product life can reasonably assess their role in lessening the overall environmental impact of their activities.

Corporate supply chains are vast, but there are many small and medium-sized companies that have little contact with these large interconnected networks. Smaller enterprises could benefit from greater exposure to the more advanced environmental practices of larger corporations. In addition, although some larger companies are at the leading edge of environmental metrics development, advancement has been far from uniform. While companies cannot be expected to release proprietary information or methods, greater efforts must be made to transfer the corporate practices more common to large manufacturers to other levels of industry.

*RECOMMENDATION 8:* **The U.S. government, acting in concert with industry, should gather and disseminate information on best practices in industrial environmental performance measurement. Improved efforts must be made to transfer the knowledge and technology of these methods across industries and sectors, particularly to small and medium-sized enterprises.**

Approaches to measuring and improving environmental performance are proliferating throughout the world. Some system needs to be devised that more effectively communicates these techniques to small and medium-sized companies as well as to larger companies that have yet to develop environmental measures. The Internet allows for the creation of a widely accessible clearinghouse of environmental metrics information. An appropriate government agency (e.g., EPA, DOC) should engage expertise from the private and public sectors to assemble and periodically update an online library of industry-specific metrics and case studies. Other avenues of dissemination might include state agencies and industry associations.

The committee recommends several different approaches to providing widespread access to information on environmental performance metrics. EPA should assume responsibility for assembling an industry- and category-specific (e.g., energy use, material use) database. Industry input will be critical to the success of such a database, in terms of identifying best practices and providing periodic updates. Furthermore, the U.S. Small Business Administration, which has offices in all 50 states, could work to increase awareness and use of environmental metrics among small and medium-sized companies.

## Goal 5: Develop Metrics that Keep Pace with New Understanding of Sustainability

Society's understanding of and commitment to the concept of sustainability is increasing. As this interest grows, all sectors of the economy must begin to investigate methods of assessing and improving the sustainability of their activities. Opportunities for cost savings, new markets, and novel methods of product differentiation are among the advantages that proactive companies might enjoy as a result of more sustainable practices. These potential benefits notwithstanding, considerable confusion still exists regarding how a company might develop and apply reasonable measures of sustainability to its activities.

While many companies have made headway in the application of eco-efficiency metrics and programs, this represent only a first step toward sustainability. As society begins to come to grips with the theoretical and technical underpinnings of sustainable development, companies will be challenged to maintain the pace of improved performance as they seek to evaluate their activities against an often changing set of criteria.

Several sustainability issues with more immediate environmental relevance include the declining natural resource base (e.g., energy and materials), alteration of global bio- and geochemical cycles (e.g., carbon, hydrologic), and threats to the world's biological base (both plant and animal). Evidence of these concerns can be seen in the goals and metrics established by certain companies. Sectors exercising some degree of control over their own resource base (e.g., paper and forestry products, fisheries, agriculture) have long recognized the value of monitoring and maintaining sustainable rates of stock growth and harvest. The potential for global warming is largely attributed to changes in the carbon cycle, prompting some companies to place a higher priority on cutting greenhouse gas emissions (and costs). Some industries have even begun to consider ways in which they might offset greenhouse emissions through the sequestration of carbon dioxide (e.g., in forests and by reinjection into spent oil and gas reservoirs). Finally, the realization that a declining biological base and the loss of genetic diversity may greatly affect advances of great importance to human beings (e.g., the cancer treatment drug taxol derived from the bark of the Pacific yew) has heightened the concerns of some companies over species loss. All of these examples are useful and directly link the profit-maximizing role of a firm with environmental improvement. Such efforts represent an important phase of society's transition to more sustainable practices. Further advances should be researched and encouraged.

*RECOMMENDATION 9:* **The U.S. government and industry should assure that adequate research attention is directed toward furthering understanding of the complex environmental interactions associated with sustainability.**

Industry's role will be driven primarily by competitive pressures as customers, investors, and regulators demand better environmental attributes from products and processes. Government's role will be to examine ways in which industrial operations and products affect the various aspects of sustainability. This may involve investigating the implications of long-term industrial activity on the environment, including such issues as materials flows and energy use.

Corporate funding may be appropriate for investigating readily defined, shorter-term topics or in cases where there is a convergence of sustainability and business objectives (e.g., the shifting emphasis of many chemical companies to biotechnology). Issues of broader scope and longer time span, however, will likely require government research support. Examples might include assessment of the implications of continued heavy dependence on landfills for hazardous and nonhazardous solid waste disposal and the resulting adverse environmental impacts on a local, regional, national, or global scale.

While the concept of sustainable development has widespread appeal, there is as yet no scientific consensus on a definition of the concept or indices by which it may be measured at the macro, or societal, level. Our knowledge of life support systems, ecosystem services, carrying capacities, eco- and social system health and integrity, and other fundamental aspects of sustainability is in its infancy. Assessing a given industrial product, process, technology, or facility with regard to sustainability will require the development of systems approaches for which very few relationships have yet been developed. Finally, while attention to purely environmental issues is important, it should be noted that economic and social concerns are integral to the concept of sustainability. Presently, however, it is difficult to directly relate a firm's contribution to many broad economic and social measures (e.g., per-capita income, average education level).

The committee recommends that the National Science Foundation (NSF), in conjunction with EPA and CEQ, initiate research programs focused on sustainable industrial enterprises. Universities will likely lead those research activities, but efforts should be made to encourage close working relationships with industry, similar to the programs under way at a number of the NSF's Engineering Research Centers (ERCs). These centers have had considerable success in working with industry, and in many cases industry has supplemented government grants with corporate funding for specific projects.

*RECOMMENDATION 10:* **Conduct research on methods of integrating socioeconomic criteria into sustainability measures.**

Research is needed to help solve the analytic, relational, and informational challenges associated with sustainability. These challenges involve not only single-issue complexities (e.g., related to the environment) but also those involving multiple behaviors and impacts (e.g., economic-environmental, social-environmental) and varying scales (e.g., local, regional, global).

Researchers must begin to examine methods of analysis and metrics that address society's ability to link environmental, economic, and social activities in a manner that can guide progress toward sustainability. It should be noted that the degree of complexity involved in these analyses may mean that useful sustainability metrics will be very difficult to devise and in some cases perhaps even impractical. Many of these metrics and the trends they identify will be far beyond industry's sphere of influence. Nonetheless, as a strong force within society, industry must begin to investigate its role in moving toward sustainability. Companies, particularly large multinationals, should begin by evaluating their products in terms of their potential for improving the physical status (e.g., related to environmental degradation), economic status (e.g., related to widespread poverty), and social status (e.g., related to urban migration) of the communities in which they operate. Although industry should play an active role, the long-term and exploratory nature of much of this research makes it more practical that nongovernmental organizations, government researchers, and the academic community take the lead in pursuing methods of quantifying these complex and interrelated aspects of sustainability.

As world population continues to rise and the Earth's carrying capacity becomes increasingly strained, sustainability issues will begin to directly impact industries and companies. If the global trend toward decentralization continues, industry will take on a more prominent role in economic and social decision making. As this responsibility and society's understanding of sustainability grow, improved methods of measuring and tracking progress toward environmentally sustainable development will become valuable tools.

The committee suggests that research into these complex interrelated topics could be most effectively overseen by the NSF. Broader and more multidisciplinary models than the successful NSF ERCs may be required to respond to this need.

## CONCLUDING REMARKS

Environmental metrics are at the heart of how industry and its many stakeholders define environmental performance and determine whether progress is being made. This report documents how the use of environmental metrics has focused the attention of companies, public agencies, and a variety of other interested parties on key areas of industrial performance. U.S. industries have integrated some metrics, especially for pollution releases and hazardous waste generation, into routine management decisions and external reporting. Many U.S. firms also track their consumption of energy and water as a basic element of cost control. Some firms have even begun developing metrics for other areas, created tools to prioritize these indicators, and experimented with more qualitative issues related to human health, ecosystem health, and broader social dimensions.

With new knowledge and changing public expectations, fresh environmental challenges are arising that are not addressed by contemporary environmental metrics. Ecosystem impacts, effects on human health, habitat loss, and global climate change are among a few of the emerging issues for which metrics are needed. To realize the full potential of environmental metrics will require changes by all the parties involved. Industry, government, and communities all have important roles to play, not only in improving industrial practices but also in extending the lessons learned in that sector to the vast array of other societal activities that impact the environment. Much work remains, but as society works to achieve a pattern of sustainable development, environmental metrics will provide a valuable tool for influencing environmental decision making and driving innovation.

## REFERENCES

Council on Environmental Quality. 1995. 25th Anniversary Report of the Council on Environmental Quality (1994–1995). Available online at http://ceq.eh.doe.gov./reports/reports.htm. [February 3, 1999]

Dow. 1996. Goals for 2005. In 1996 Environment, Health, and Safety Report. Available online at http://www.dow.com/cgi-bin/frameup.cgi?http://www.dow.com/environment/goal2005.html. [February 3, 1999]

United States Environmental Protection Agency (USEPA). 1987. Unfinished Business. Office of Policy Analysis. Washington, D.C.: USEPA.

United States Environmental Protection Agency (USEPA). 1990. Reducing Risk. Science Advisory Board. Washington, D.C.: USEPA.

United States Environmental Protection Agency (USEPA). 1992. National Air Quality and Emissions Trends. EPA-450-R-92-001. Office of Air Quality Planning and Standards. Research Triangle Park, N.C.: USEPA.

United States Environmental Protection Agency (USEPA). 1998. Environmental Quality, Status, and Trends. Available online at http://www.epa.gov/ceis. [February 3, 1999]

# APPENDIXES

APPENDIX A
# Current Reporting and Use of Industrial Environmental Performance Metrics: Global Scale

Recent years have seen a proliferation in both the amount and type of environmental reporting done by industry. A 1997 report by KPMG summarizes reporting practices by over 900 firms from 13 OECD (Organization for Economic Cooperation and Development) countries (Figures A-1, A-2). Seventy-one percent of companies surveyed made some mention of the environment in their annual report, up from 58 percent in 1993. The proportion of companies producing separate environmental reports rose from 15 percent in 1993 to 24 percent (220) in 1997 (Figure A-3). Forty-one percent of U.S. companies surveyed reported releasing a separate environmental report (Figure A-4).

The topics covered by industry environmental reports were generally weighted toward the traditional concerns of emissions and compliance (Figure A-5). However, some movement toward ecoefficiency was evident in terms of the emphasis some firms gave to natural resource conservation. In a further indication of more visionary behavior, the concept of sustainable development was mentioned in some reports. Increased attention given to such issues as waste management, energy conservation, supplier performance, and product design when discussing future goals indicates that a number of companies have begun to move beyond compliance as the sole motivation for improving environmental performance (Figure A-6). While all of these topics were mentioned in company reports, quantifiable measures of performance in these areas are still somewhat sparse.

Of those companies producing separate environmental reports, 87 percent (192) disclosed quantitative environmental performance data. Disclosures of most such data were weighted primarily toward emissions, but attention was also

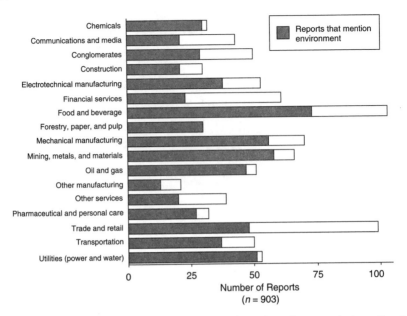

FIGURE A-1  Total number of annual reports and number of reports that mention the environment, by industry sector. SOURCE: KPMG (1997).

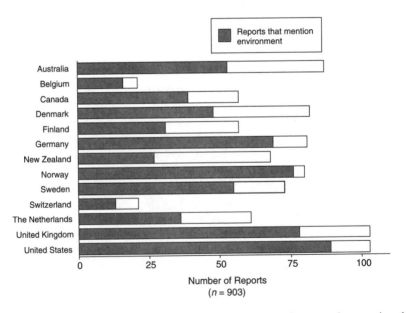

FIGURE A-2  Total number of annual reports and number of reports that mention the environment, by country. SOURCE: KPMG (1997).

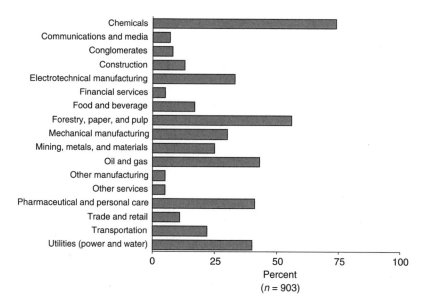

FIGURE A-3  Percent of companies producing separate environmental reports, by industry sector.  SOURCE: KPMG (1997).

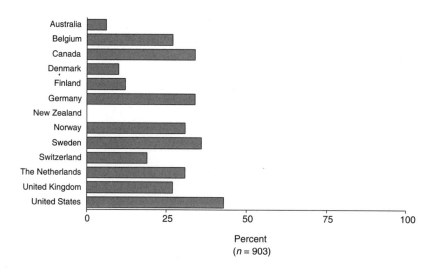

FIGURE A-4  Percent of companies producing separate environmental reports, by country.  SOURCE: KPMG (1997).

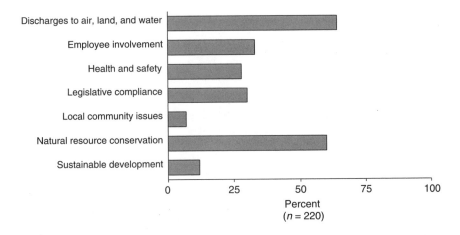

FIGURE A-5  Content of company environmental reports that address specific environmental topics.  SOURCE: KPMG (1997).

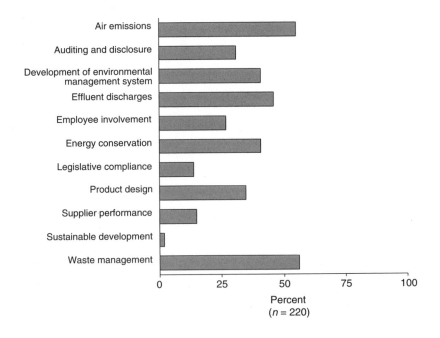

FIGURE A-6  Percent of environmental reports that discuss future plans and targets, by topic.  SOURCE: KPMG (1997).

given to assessing environmental costs and conservation efforts (Figure A-7). The disclosure of quantitative information by international respondents in the KPMG survey can be compared with that of U.S. firms in a smaller study (Figure A-8). The attention given to environmental expenditures in both surveys demonstrates a growing awareness of these costs and a move away from the common practice of simply lumping them in with overhead expenses. Such awareness may indicate the first steps toward more comprehensive assessment of both the costs and benefits associated with environmental programs and capital investments. It is also interesting to note that almost half (42 percent) of U.S. companies surveyed by KPMG produced environmental reports that included quantitative

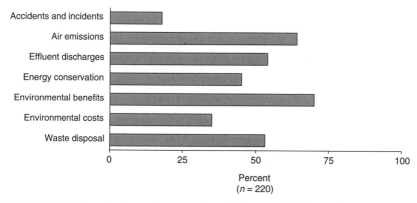

FIGURE A-7  Percent of annual reports that disclose quantitative environmental data, by type.  SOURCE: KPMG (1997).

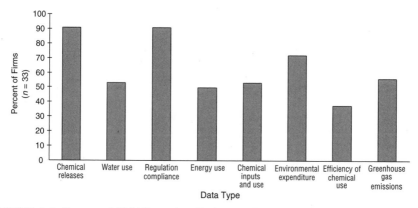

FIGURE A-8  Percent of U.S. firms that disclose environmental performance data, by type of data.  SOURCE: White and Zinkl (1996).

targets and deadlines with regard to improving environmental performance (Figure A-9). It is apparent that some companies are approaching the use of environmental metrics with vigor, particularly in areas important to tracking internal performance. However, relatively few measures of ecosystem impacts or the sustainability of industrial activities were noted in either survey.

Environmental management issues were mentioned in a significant percentage of environmental reports (Figure A-10). Although often lacking quantifiable data, many of the reports contained information on such matters as management responsibility for the environment and corporate environmental management systems. It is perhaps a sign of progress that a significant fraction (43 percent) of those companies producing environmental reports included details of internal and third-party environmental audits. While those doing so represented only about 10 percent of total survey respondents, the release of such information in years past would have been unlikely.

The number of issues and the amount of quantifiable data being reported are increasing, and while a number of different approaches to standardizing environmental metrics and reporting have been proposed, little consensus has emerged.

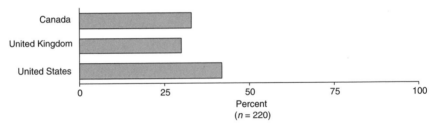

FIGURE A-9  Percent of company environmental reports that include quantitative targets, deadlines, and reporting on targets, by country.  SOURCE:  KPMG (1997).

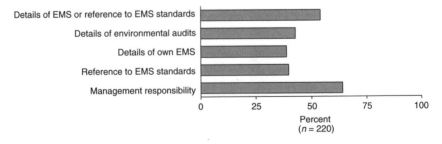

FIGURE A-10  Percent of environmental reports that contain environmental management system (EMS) information, by type.  SOURCE:  KPMG (1997).

There are many complexities associated with establishing a code of internal metrics for corporate use, and they vary greatly by industry. A review of present practices reveals that industry has broad opportunities to improve on the current slate of environmental performance metrics.

## REFERENCES

KPMG. 1997. International Survey of Environmental Reporting. Lund, Sweden: KPMG.

White, A., and D. Zinkl. 1996. Corporate Environmental Performance Indicators: A Benchmark Survey of Business Decision Makers. Boston: Tellus Institute.

# Current Reporting of Industrial Environmental Performance Metrics by U.S. Semiconductor Companies

To examine the comparability of environmental metrics reported (and presumably used by companies) within an industry, the committee surveyed the annual and environmental reports of semiconductor manufacturers (also found on the websites of the companies). Both the categories and units of measurements used by these companies are shown below. Table B-1 summarizes the categories of metrics covered by this small set of companies.

**Intel Corporation (http://www.intel.com/intel/other/ehs/index.htm)**

VOC (volatile organic compound) emissions (tons per year)
Hazardous organic and inorganic air pollutant emissions (tons)
$NO_x$, CO, and $SO_2$ emissions (tons per year)
Phospene emissions (tons)
Sulfuric acid emissions (tons)
Particulate emissions—particulate matter measuring 0–10 μm (tons)
Hazardous waste generation (thousand tons per year, recycled)
Solid waste recycling—cardboard, metal, paper, plastic, wood (percent recycled, tons per year), other
Chemical waste management—incineration, other treatment or disposal, recycling, energy recovery (percent)
Nonhazardous chemical waste management (tons, recycled)
Water use (millions of gallons per year, gallons per employee per day)
Wastewater reuse (total gallons and percent reuse)
Energy use (million kilowatt-hours)

Natural gas (million cubic feet)
Superfund Amendments and Reauthorization Act (SARA) Title III releases and
off-site treatment (tons per year)
SARA Title III reportable chemicals by site—releases to air, transfers off-site
(pounds)
Compliance issues (violations, fines, corrective actions)
Regulatory inspections and citations (number/$)
Award and recognition

## Texas Instruments (http://www.ti.com/corp/docs/esh/index.htm)

Hazardous waste—disposal, recycled, recycle off-the-shelf chemicals,
manufacturing waste (1,000s of pounds, percent, and manufacturing
output)
Nonhazardous waste (1,000s of pounds, percent disposal, and recycled)
33/50 chemicals (percent of total used and total released to air, surface water;
United States only)
SARA chemicals (1,000s of pounds and percent of total used that is released to
air, land, and water; United States only)
Greenhouse gases (pounds of PFCs [perfluorocarbons] released or percent of
total used)
Energy usage—electricity, gas, fuel, water (BTUs)
Recycled packaging
Inspections, fines, and violations
Awards and recognition

## Motorola Corporation (http://www.mot.com/EHS/)

Hazardous waste—disposal, recycled, manufacturing waste (quantity and
production unit)
Nonhazardous waste—disposal, recycled (quantity/production unit)
VOC air emissions (tons)
Toxic Release Inventory (TRI) releases (millions of pounds)
33/50 chemicals (percent of total used and pounds that are released to air,
surface water; United States only)
SARA chemicals (total 1,000s of pounds and percent of total used that is
released to air, land and water; United States only)
Energy usage—electricity, gas, fuel (BTUs)
Water usage (gallons)
Awards
Inspections, fines, and violations

## Hewlett Packard (http://www.hp.com/abouthp/envrnmnt/)

Hazardous waste generation and disposal (millions of pounds)
Nonhazardous solid waste recycling and disposal (millions of pounds)
ODSs (ozone-depleting substances; pounds)
TRI Chemicals (millions of pounds)
33/50 Chemicals (millions of pounds)
Global warming gases (PFC emissions)
Product/shipping packaging
Energy consumption and product use

## National Semiconductor (http://www.national.com/environment/)

Hazardous waste
Nonhazardous waste
Electronic scrap
Chemical emissions (pounds/$1,000 sales)
33/50 Chemicals (total released in 1,000s of pounds)
Awards and recognition
ODSs
Water
Energy

## Lucent Technologies (http://www.lucent.com/environment/)

Energy (millions of dollars in savings)
Air emissions
CFCs (chlorofluorocarbons)
VOCs
Hazardous waste disposal (millions of pounds)
Nonhazardous solid waste—recycling, disposal (millions of pounds)
ESH (environment, safety, and health) audits and corrective actions
Environmentally responsible manufacturing
Environmentally responsible packaging
Greenhouse gas emissions
Water usage
Remediation
Compliance
Awards and recognition

## AMD  (http://www.amd.com/about/investor/1997annual/annual.html)

Hazardous waste (1,000s of pounds)

Nonhazardous solid waste (1,000s of pounds)
TRI emissions (1,000s of pounds, percent)
PFCs
Energy
Water
Packaging
Awards and recognition


**Digital Equipment  (http://www.digital.com/ehs/)**

Hazardous waste—office, nonoffice (kilograms)
Nonhazardous solid waste (kilograms)
Compliance (notices of violation, penalties)
Energy consumption (million kilowatts-hours, kilowatt-hours/square foot)
Remediation
Air emissions (kilograms)
Water emissions (kilograms)
Toxic chemical use in production (kilograms)
VOC emissions (kilograms)
ODS usage—total CFC, hydrofluorocarbons, halon emitted (kilograms)
Total water use (millions of cubic meters)
Water use per employee (millions of cubic meters)


**International Business Machines (http://www.ibm.com/ibm/environment/)**

Hazardous waste reduction—closed-loop recycling, off-site recycling,
    treatment, and disposal (percent, tons)
Nonhazardous waste recycled (percent recycled, total generated)
PCB (polychlorinated biphenyl) waste
33/50 Chemicals usage (percent)
SARA Title III releases and transfers (percent, tons)
Ozone concentration in California (parts per billion)
Energy (kilowatt-hours, gallons of fuel saved, tons of carbon dioxide emissions)
Environmental costs and savings
PFCs
Water (gallons/year)
Spills and releases
Fines and penalties
Recycled materials in new products (percent)
Recycled plastic use in new products (percent)
Landfill space used for IBM equipment dismantling operations (percent)
Awards and recognition

**Rockwell International (http://www.rockwell.com/about/env/)**

Hazardous waste generation (1,000s of tons, disposal type)
SARA emissions, releases, and transfers (millions of pounds, percent)
33/50 Emissions (millions of pounds, 1,000s of pounds, percent by chemical)
ODS chemicals (percent)
Fuel and energy usage
Carbon dioxide emissions (1,000s of tons)
Environmental costs
Environmental compliance (fines, violations, penalties)
Awards and recognition

TABLE B-1 Data Collected from Company Public ESH Annual Reports

| | Intel | Texas Instruments | Motorola | Hewlett Packard | National Semiconductor | Lucent | AMD | Digital Equipment | IBM | Rockwell |
|---|---|---|---|---|---|---|---|---|---|---|
| Environmental costs | | | | | | | | | | • |
| Landfill space used for dismantled equipment | | | | | | | | | • | |
| Recycled plastic used in new products | | | | | | | | | • | |
| Recycled materials in new products | | | | | | | | | • | |
| Polychlorinated biphenyls | | | | | | | | | • | |
| Toxic chemicals use in production | | | | | | | | • | | |
| Remediation | | | | | | | | • | | |
| Environmentally responsible manufacturing | | | | | | | • | | | |
| Perfluorocarbons | | | | | | | • | | | |
| Air emissions | | | | | | | • | • | | |
| Chemical emissions | | | | | | • | | | | |
| Electronic scrap | | | | | | • | | | | |
| Global warming | | | | • | | | | | | |
| Ozone depleting substances | | | | • | • | • | | • | | • |
| Packaging | | • | | • | | • | • | | | |
| Greenhouse gases | | • | | | | • | | | | |
| 33/50 chemicals | | • | • | • | • | | | | | • |
| Awards and recognition | • | • | • | | | • | • | | • | • |
| Regulatory inspections | • | • | • | | | | • | • | • | • |
| Compliance issues | • | • | • | | | | • | • | • | • |
| SARA/TRI chemicals | • | • | • | • | | | • | | | • |
| Natural gas | • | • | • | • | • | • | • | • | | |
| Fuel use | • | | | | | | | | | • |
| Energy Use | • | • | • | • | • | • | • | • | | • |
| Water use in manufacturing | | | | | | | | • | | |
| Water use per employee | | | | | | | | • | | |
| Water use | • | • | | | • | • | | • | | |
| Chemical waste management | • | | | | | | | | | |
| Nonhazardous | • | • | • | • | • | • | • | • | • | |
| Hazardous waste | • | • | • | • | • | • | • | • | • | • |
| $NO_x/CO_2$ emissions | • | | | | | | | | | • |
| Volatile organic chemicals | • | | | | | | • | • | | |

NOTE: Highlighted column entries represent metrics tracked and reported by four or more companies.

# APPENDIX C
# Current Reporting and Use of Industrial Environmental Performance Metrics: Company Scale[1]

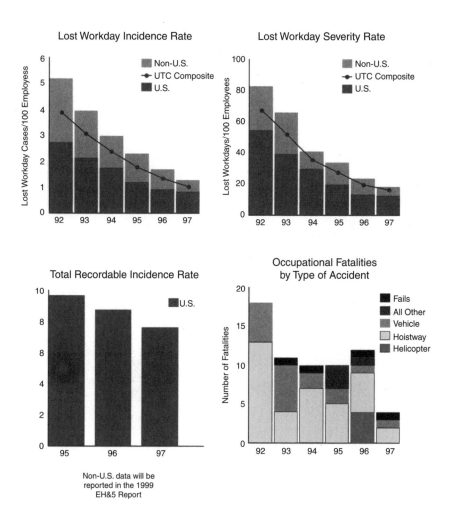

### Lost Workday Incidence Rate

### Lost Workday Severity Rate

### Total Recordable Incidence Rate

Non-U.S. data will be
reported in the 1999
EH&S Report

### Occupational Fatalities by Type of Accident

---

[1]Data for all figures in Appendix C from *Environment, Health, and Safety Report,* United Technologies Corporation, 1997.

### U.S. Chemical Releases

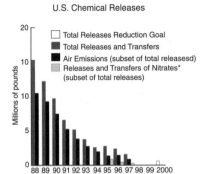

Legend:
- Total Releases Reduction Goal
- Total Releases and Transfers
- Air Emissions (subset of total releasesd)
- Releases and Transfers of Nitrates* (subset of total releases)

*Total releases of water dissociable nitrates were reportable beginning in 1995

### Total U.S. Hazardous Waste Generated

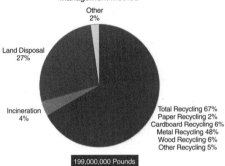

Legend:
- Actuals
- Goal

### U.S. Hazardous Waste Sent Off Site by Process Source

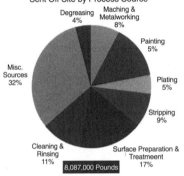

Degreasing 4%
Maching & Metalworking 8%
Painting 5%
Plating 5%
Stripping 9%
Surface Preparation & Treatmeent 17%
Cleaning & Rinsing 11%
Misc. Sources 32%

8,087,000 Pounds

### U.S. Solid Waste Management method

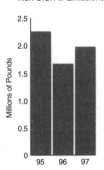

Other 2%
Land Disposal 27%
Incineration 4%

Total Recycling 67%
Paper Recycling 2%
Cardboard Recycling 6%
Metal Recycling 48%
Wood Recycling 6%
Other Recycling 5%

199,000,000 Pounds

### Non-U.S. Industrial Waste Generation

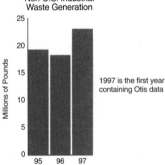

1997 is the first year containing Otis data

### Non-U.S. Air Emissions

U.S. Releases and Transfers
Ozone Depleting Substances

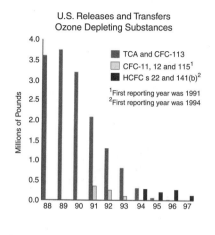

Worldwide Energy Consumption
Fuel Type by Btu Equivalence

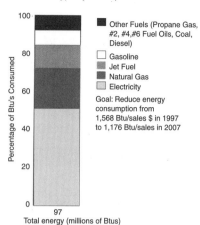

Worldwide Water CO$_2$ Emissions
by Fuel Type (Metric Tons)

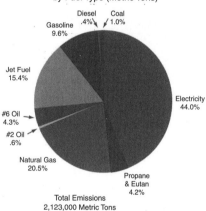

Worldwide Water Consumption

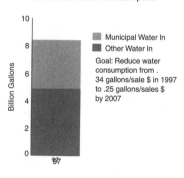

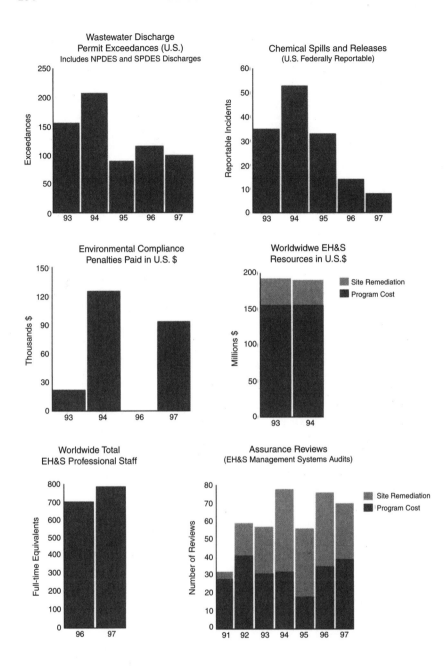

Wastewater Discharge
Permit Exceedances (U.S.)
Includes NPDES and SPDES Discharges

Chemical Spills and Releases
(U.S. Federally Reportable)

Environmental Compliance
Penalties Paid in U.S. $

Worldwidwe EH&S
Resources in U.S.$

Worldwide Total
EH&S Professional Staff

Assurance Reviews
(EH&S Management Systems Audits)

# Biographical Data

**ROBERT A. FROSCH** (*Chair*) is a senior research fellow at the Center for Science and International Affairs of the John F. Kennedy School of Government, Harvard University, and a senior fellow at the National Academy of Engineering. In 1989 he revived, redefined, and popularized the term "industrial ecology," and his researched has focused on this field in recent years, especially in metals-handling industries. Frosch retired as vice president of General Motors Corporation in 1993, where he was in charge of the North American Operations Research and Development Center. He has been involved in global environmental research and policy issues at both the national and the international levels. Frosch's career combines varied research and administrative experience in industry and government service, including positions as deputy director of the Advanced Research Projects Agency of the U.S. Department of Defense, assistant secretary of the Navy for research and development, assistant executive director of the United Nations Environment Programme, and administrator of the National Aeronautics and Space Administration. He has a Ph.D. in theoretical physics from Columbia University and is a member of the National Academy of Engineering.

**DAVID C. BONNER** is director of technology, Performance Polymers Division, Rohm and Hass. He was previously senior vice president, technology and engineering, for the Westlake Group. Before joining Westlake in 1996, he was senior vice president and chief technical officer of Premix; vice president, research and development, at BFGoodrich; and associate professor of chemical engineering at Texas A&M University. Bonner has published over 50 articles in the field of chemical engineering in peer-reviewed journals. He is a member of the American Institute of Chemical Engineering and the American Chemical Society. Bonner has a Ph.D. in chemical engineering from the University of California at Berkeley.

**JOHN B. CARBERRY** is director of environmental technology for E.I. DuPont. He has responsibility for research and development that is focused on process renewal, product stewardship and product recycle. Carberry's responsibilities include providing technical assistance in understanding emerging environmental issues and analysis of their impact on DuPont, and reviewing the opportunities for new business based on environmental excellence. In addition, he is responsible for a corporate team that leads initiatives to obtain world-class, affordable, publicly acceptable remediation, treatment, and abatement technologies. Since 1988, Carberry has helped oversee a transition to waste prevention at DuPont, while maintaining excellence in the company's treatment, abatement, and remediation efforts. He is presently serving as chairman of the chemical engineering advisory board at Cornell University, is on the radioactive waste retrieval Technology Review Group for the U.S. Department of Energy, and sits on the American Chemical Society's Pollution Prevention Program Committee. Carberry has a master's degree in chemical engineering from Cornell University and an M.B.A. from the University of Delaware.

**LESLIE CAROTHERS** is vice president, environment, health, and safety at United Technologies Corporation (UTC) and is an adjunct lecturer on environmental regulation at the Yale University School of Forestry and Environmental Studies. She has been an environmental professional for 26 years. Before joining UTC in 1991, Carothers served as commissioner of the Connecticut Department of Environmental Protection from 1987 to 1991. She is chair of the board of directors of the Connecticut Audubon Society, vice chair of Leadership Greater Hartford, and a past board member of the Environmental Law Institute. During 1994–1995, Carothers served on the Defense Science Board's Environmental Security Task Force, which was assigned to make recommendations for improvements in the environmental programs of the U.S. Department of Defense. She is a graduate of Smith College and Harvard Law School and has a master's degree in environmental law from George Washington University.

**DARYL DITZ** is director of environmental management programs at the Environmental Law Institute (ELI). He is leading a new project on sustainable development in the southern United States and extending ELI's programs on corporate environmental management and public policy in Asia. Between 1993 and 1997, Ditz led several projects at the World Resources Institute, including innovative work on corporate environmental accounting presented in the book *Green Ledgers: Case Studies in Corporate Environmental Reporting* (World Resources Institute, 1995) and extensive work with the electronics industry and the U.S.

forestry sector. He has lectured and published widely on risk management, information disclosure, and other aspects of environmental policy. Ditz has a Ph.D. in engineering and public policy from Carnegie Mellon University.

**THOMAS N. GLADWIN** is the Max McGraw Professor of Sustainable Enterprise and director of the Corporate Environmental Management Program, a joint initiative of the Business School and School of Natural Resources and Environment at the University of Michigan. His research and teaching focus on ecologically and socially sustainable business on a global basis. Gladwin has taught multinational corporate executive programs and has served as a consultant to the Corporate Conservation Council of the National Wildlife Federation, Business Council for Sustainable Development, U.S. Environmental Protection Agency, Organization for Economic Cooperation and Development Environment Directorate, National Science Foundation, and World Commission on Environment and Development. He has a Ph.D. in international business and natural resource policy from the University of Michigan.

**THOMAS E. GRAEDEL** is a professor of industrial ecology, a professor of chemical engineering, and a professor of geology in the School of Forestry and Environmental Studies at Yale University, a position he assumed after 27 years at AT&T Bell Laboratories. He was the first atmospheric chemist to study the atmospheric reactions of sulfur and the concentration trends in methane and carbon monoxide. As a corrosion scientist, he devised the first computer model to simulate the atmospheric corrosion of metals. This work led to a voluntary position as consultant to the Statue of Liberty Restoration Project in 1984–1986. One of the founders of the newly emerging field of industrial ecology, Graedel coauthored the first textbook in that specialty and has lectured widely on its implementation and implications. He has published nine books and more than 200 scientific papers. He has a B.S. from Washington University, an M.A. from Kent State University, and an M.S. and a Ph.D. from the University of Michigan.

**CHRISTOPHER (KIT) GREEN** is executive director of the materials research and technology business development directorate and chief technology officer, China, for General Motors Corporation. The directorate focuses on advancing research for materials utilizing physics and chemistry as core disciplines. As GM's chief technology officer in China, Green manages and coordinates technology acquisition and deployment for corporate joint ventures. He is a diplomat of the Federation of State Medical Boards of the United States, a medical officer in the U.S. government, and a recipient of the National Intelligence Medal. Green is the former chairman of the National Research Council (NRC) Committee on Biotechnology in the Year 2020. He is chairman of the NRC Board on Army Science and Technology and a member of numerous National Academy of Sciences and Institute of Medicine commissions. Green has a Ph.D. in neurophysiology from the University of Colorado Medical School and an M.D. from the University Autonoma in El Paso, Texas/Juarez.

**RICHARD R. GUSTAFSON** is the Denman Professor of Paper Science and Engineering and chair of the Management and Engineering Division at the University of Washington College of Forest Resources. His research is focused on the analysis, modeling, and control of pulping and bleaching operations, and the department he chairs covers the entire range of the forestry of pulp fibers. Gustafson has done research and has consulted for several forest products companies, including Weyerhaeuser, Potlatch, and Champion International. He has a B.S. in wood and fiber science and a Ph.D. in chemical engineering from the University of Washington.

**MICHAEL J. LEAKE** is director of environment, health, and safety for Raytheon/Texas Instruments Systems. Previously, he was quality assurance manager of group manufacturing and was responsible for metal fabrication, printed-wiring board manufacturing, magnetics, and plastics manufacturing. Leake's research interests are in environmental and thick-film engineering. He is cochairman of the Environmental Enhancement Group/21st Century Manufacturing Enterprise Strategy, which is tasked with assessing U.S. industrial competitiveness and determining what is needed to maintain worldwide competitiveness to the year 2006. Leake is also a member of the Steering Committee for the Environmentally Conscious Manufacturing Strategic Initiative. He has an M.S. in environmental science from the University of Texas at Dallas.

**DAVID W. MAYER** is director of pollution prevention and environmental performance at Georgia-Pacific and director of the Global Environmental Management Initiative. He is responsible for implementing Georgia-Pacific's 58 Environmental Principles and Goals, as well as developing future environmental performance measures. In addition, Mayer develops an annual company-wide strategy to provide environmental training for officers, key managers, and environmental coordinators. Previously, he was chief of the technical assistance staff for the U.S. Environmental Protection Agency (EPA) Asbestos Action Program, where he coordinated and reviewed the development of all of EPA's guidance documents on asbestos. Subsequently, Mayer became director of the Environmental Training Center for Law Environmental and was responsible for coordinating the development, staffing, and implementation of a wide range of environmental health and safety training courses. He has a B.S. in biology and environmental studies from Coe College and an M.S. in engineering and public policy from Washington University.

**RICHARD D. MORGENSTERN** is a senior fellow at Resources for the Future, on leave from the U.S. Environmental Protection Agency (EPA) and is currently engaged as a senior economic counsellor at the U.S. Department of State. He was a tenured associate professor of economics at Queens College of the City University of New York before becoming deputy assistant director for energy, the environment, and natural resources at the Congressional Budget Office. Subsequently, Morgenstern served as legislative assistant to Sen. J. Bennett Johnston and then directed the energy program of the Urban Institute. He later directed the office of policy analysis of the EPA. At EPA, Morgenstern has served as acting assistant administrator for policy, planning, and evaluation (1991–1993) and as deputy administrator during the transition period at the beginning of the Clinton administration. He has a Ph.D. in economics from the University of Michigan.

**WILLIAM F. POWERS** is vice president of research at Ford Motor Company. Prior to this assignment he was executive director of information technology at the Ford Research Laboratory. Before joining Ford in 1979, Powers was professor of aerospace engineering and computer, information, and control engineering at the University of Michigan. He serves as a member of the USCAR Council, the Automotive Research Center Board of Advisors, and the NRC's Competitiveness Task Force. Powers is a member of the National Academy of Engineering and the American Society of Mechanical Engineers, a foreign member of the Royal Swedish Academy, and a fellow of the Institute of Electrical and Electronics Engineers. He has a Ph.D. in engineering mechanics from the University of Texas at Austin.

**DARRYL K. WILLIAMS** is senior vice president, technology and environment, for Eastman Chemical Company. He joined Eastman Chemical in 1965 and has served in a variety of manufacturing management positions. Williams joined Eastman Kodak in 1990 as manager of strategic planning for consumer color film. He returned to Eastman Chemical in 1992 as president of Eastman Chemical Japan Ltd. Williams was appointed vice president, Asia Pacific regional support services, in 1993 and vice president, Asia Pacific sales, in 1994. He has a master's degree in chemical engineering from the University of Tennessee and was a Sloan fellow at the Massachusetts Institute of Technology where he received a master's degree in management.

## COMMITTEE STAFF

**DEANNA J. RICHARDS** is associate director of the National Academy of Engineering's program office and directs the Academy's program on Technology and Sustainable Development, formerly known as Technology and Environment (T&E). Hired in 1991 to launch the T&E program, projects, publications, and dissemination efforts under her direction have helped catalyze the establishment of the field of industrial ecology. Richards's efforts have focused on technological innovation and management issues critical to meeting environmental goals.

The approach has used analysis of the flows of materials energy, labor, and capital to support smarter strategies and decision making in industrial firms. Another focus has been to understand how technological innovations help firms, nations, and entire regions adapt to economic opportunities and environmental constraints. An environmental engineer by training, she has a B.S. (with honors) in civil engineering from the University of Edinburgh in Scotland and an M.S. and a Ph.D. from the University of Pennsylvania.

**GREGORY W. CHARACKLIS** was selected as a National Academy of Engineering fellow in 1997 to work on issues related to industrial environmental performance metrics. He hails from the Department of Environmental Sciences and Engineering at Rice University, where his research focused on applying market mechanisms to water management in Texas. Prior to moving to Rice University in 1992, Characklis was associate scientist in the research division of EG&E, Inc., at the Idaho National Engineering Laboratory, a U.S. Department of Energy facility. He has a bachelor's degree in materials science and engineering from Johns Hopkins University and master's and doctoral degrees in environmental engineering from Rice University.

**GREG PEARSON** is editor of the National Academy of Engineering. He has over 15 years of experience as an editor and writer in technical and policy aspects of science, health, and technology. Prior to joining the NAE in 1995, Pearson was associate editor at *Science News* magazine. Before joining the *Science News* staff in 1993, he ran a successful freelance editing and writing business, consulting on various projects for the National Academy of Sciences, National Research Council, Institute of Medicine, National Institutes of Health, and National Science Foundation. Pearson has an M.A. in journalism from the American University and a B.A. in biology from Swarthmore College.

**LONG T. NGUYEN** joined the National Academy of Engineering in 1997 as a project assistant, following a personal three-month travel-study mission to Vietnam and assignments with the Commission on Physical Sciences, Mathematics, and Applications, and the Computer Science and Telecommunications Board, both at the National Research Council. He has a B.S. in international economics from Georgetown University.

# Index